LUCY SUMMERS runs her own successful landscape design partnership, the Open Garden Company, which has both national and international clients. As an RHS-qualified horticulturalist she has staged show gardens at the Chelsea Flower Show and has been awarded much-coveted Gold and Silver medals for her garden designs. She also contributes regularly to gardening publications, gives lectures to gardening clubs and organisations, and co-hosted *Britain's Best Back Gardens* for ITV among other television work. She lives in Surrey.

66 Trying to find a gardening book that is relevant to your needs isn't as straight-forward as it seems. Whether you are new to gardening or a dab hand, sometimes plant descriptions, including care, maintenance and general gardening jargon, can seem overly complicated or, worse still, just too vague. Greenfingers Guides cut through all this, delivering honest, practical information on a wide variety of beautiful plants with easy-to-follow layouts, all designed to enable you to get the best from your garden. Happy gardening! 99

GREENFINGERS GUIDES

DROUGHT-TOLERANT PLANTS

LUCY SUMMERS

headline

Copyright © Hort Couture 2009
Photographs © Garden World Images Ltd
except *Rosa* Summer Song (p. 28) © David Austin Roses
and *Agapanthus* 'Black Pantha' (p. 46) © R. Fulcher, Pine
Cottage Plants

The right of Lucy Summers to be identified as the Author
of the Work has been asserted by her in accordance with
the Copyright, Designs and Patents Act 1988.

First published in 2009
by HEADLINE PUBLISHING GROUP

1

Lucy Summers would be happy to hear from readers
with their comments on the book at the following
e-mail address: lucy@greenfingersguides.co.uk

The Greenfingers Guides series concept was originated
by Lucy Summers and Darley Anderson

A CIP catalogue record for this title is available from
the British Library

ISBN 978 0 7553 1759 2

Design by Isobel Gillan
Printed and bound in Italy by Canale & C.S.p.A.

Headline's policy is to use papers that are natural,
renewable and recyclable products and made from wood
grown in sustainable forests. The logging and
manufacturing processes are expected to conform to the
environmental regulations of the country of origin.

HEADLINE PUBLISHING GROUP
An Hachette Livre UK Company
338 Euston Road
London NW1 3BH

www.headline.co.uk
www.hachettelivre.co.uk
www.greenfingersguides.co.uk
www.theopengardencompany.co.uk

PICTURE CREDITS
All photographs supplied by Garden World Images

ACKNOWLEDGEMENTS
My thanks to Darling, Zoe, Serena, Lorraine, Emma, Josh,
Isobel, Charlotte, and Michael Loftus of Woottens Plants.
And to all my wonderful nearest and dearest.

OTHER TITLES IN THE GREENFINGERS GUIDES SERIES:

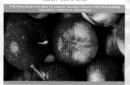

Climbers and Wall Shrubs
ISBN 978 0 7553 1758 5

Fruit and Vegetables
ISBN 978 0 7553 1761 5

Contents

Introduction

For years I gardened on heavy clay in East Anglia, which has one of the lowest rainfalls in the UK, and a long period of drought meant potential ruin for many areas of my garden. Large cracks opened up in the dry soil and plants keeled over from heat exhaustion. It was more through necessity than choice that I began employing drought-combating measures in the garden, having witnessed far too many plant fatalities over a long, hot summer. There was insufficient water to cope with all the plants crying out for a drink, let alone the time to water them all if I had the water in the first place. Even in the UK, with the typical British lamentations over our mercurial weather, there are times when a gardener prays for rain!

Mulching was my first course of action, and as my acreage was fairly large I shovelled on both garden and mushroom compost by the lorry load. What a difference it made! My watering regime was cut in half, leaving more time for other garden tasks.

Then I started reading up on plants that were naturally drought tolerant and visited well-known 'dry' gardens, including Beth Chatto's innovative gravel garden which happened to be near by, to beg, steal and borrow any ideas, from drought-cultivation techniques to plant choices I could implement successfully in my own garden. This meant choosing plants that were capable of with-standing prolonged dry weather, instead of being seduced by the more showy, leafy herb-aceous plants that were so typical in my borders. This was hardly hard work as plant sowing and shopping are most agreeable pastimes, and drought lovers have all the charm any plant could want, offering a wealth of colour, fragrance, gorgeous foliage and, in many cases, superb architectural form.

ABOVE *Euphorbia myrsinites*   OPPOSITE *Verbena bonariensis*

Over a three-year period, I steadily altered the way I gardened; many plants needed less of my attention, seeping hoses took care of the watering when it was needed for the less drought-tolerant plants, and my fruit garden, which was way past its best, was dug up to make way for a Mediterranean garden that thrived on little or no water and hardly any maintenance.

I hope this book will encourage you to adopt a different way of gardening and excite you into growing some beautiful plants that you may never have considered before: they will give sturdier service than many of the more traditional herbaceous perennials that swoon resignedly at the very thought of water deprivation. When the dreaded hosepipe bans kick off, drought-friendly garden plants will not only survive the dry conditions, they will positively embrace long, hot summers, displaying their full finery unimpaired.

Choosing drought-tolerant plants

Let's get straight to the heart of the matter. What is a drought-tolerant plant? All plants need some form of moisture to survive, but plants that are described as drought tolerant can come through a prolonged period of low rainfall or water scarcity without suffering lasting harm or death. Many naturally drought-tolerant plants come from areas that suffer lengthy periods of little or no rain, such as the Mediterranean. They are used to full sun and poor soils and don't struggle with dry conditions.

Choosing the right plant for the right place is of utmost importance. This is a central principle of drought gardening. Put sun lovers in the sun and shade lovers in the shade. Obvious though it may seem, I can't stress this point strongly enough: you would be surprised how many gardeners stick their plants in any old place, cross their fingers and hope for the best!

I am very much of the opinion that many herbaceous plants get a bit spoilt when they are grown in our temperate climate, often in soil that is richer than their indigenous soils and benefiting from more rainfall than they are used to. Placing them somewhere that more or less mimics the area they are from provides them with an ideal environment and diet on which they can thrive.

Water scarcity impairs a plant's vigour, so it is unable to develop as robustly as when water is plentiful. This is especially true of the more leafy herbaceous perennials, which do not do well in drought: large leaves lose more water, more quickly. However, there are some familiar perennials which are not usually thought of as drought tolerant (among them acanthus and aquilegia) whose ability to cope in dry conditions can be boosted by drought-combating cultivation methods, such as improving the soil and mulching generously.

The look and needs of drought-tolerant plants are both going to affect the way they are used in the garden. Drought lovers often have an interesting, architectural habit, and silver or glaucous foliage is common. These plants have inspired a completely different approach to garden design, as found in Mediterranean and gravel gardens as well as wildflower or prairie planting schemes. An easy starting point is to give lush, herbaceous borders a makeover, substituting drought-tolerant perennials for some of the more greedy plants. Your garden will immediately have a more modern interpretation as well as being lower maintenance in the long run.

The plants listed in this book are those that will perform well in drought conditions in this country, based on personal experience or recommendations from other gardeners. The plants have been selected on the assumption that they are established, mature plants with good root development and will be able to do without water for a reasonable length of time, but we are talking about a matter of weeks here, not months on end.

Drought tolerance is not an exact science: the survival of particular plants will depend on the soil conditions and growing environment, and the general health and age of the plant, together with the judgement and skill of the individual gardener. Additionally, the plant suggestions in this book are by no means exhaustive. For reasons of space, trees (with the exception of the olive), annuals and bedding plants have been excluded, but I have included many perennials, plus some of the smaller shrubs that are so precious in formulating an attractive drought-tolerant planting scheme.

Many drought-proof plants will be no strangers to you: lavenders and rosemary are widely grown because we love them so. The fact they keep on trooping through dry weather is just an unexpected bonus.

Adopting a drought-gardening philosophy will change the way you garden, from cultivation techniques to plant choices. Perhaps it will also encourage you to consider fresh design concepts. Gardening is fun, rewarding and challenging: embracing drought gardening is a positive, creative way of meeting the challenges wrought by climate change while giving us renewed enthusiasm for the gardens we love so well.

The bold colour of the Peruvian lily (*Alstroemeria aurea*)

Using this book

Each plant listed is categorised according to season and its eventual height, with useful, practical cultivation advice that will encourage you to grow with ever greater enjoyment, creativity and confidence. More detailed information, covering all the different elements mentioned in the profiles, and including help with planting and propagation, will be found after the plant profiles. Lists of plants for specific purposes can be found at the back of the book.

Throughout the book, plants are arranged seasonally, but in practice the corresponding months will vary according to local weather patterns, regional differences and the effects of climate change. Additionally, the flowering times of many plants span more than one season. The seasons given are based in this country, and should be thought of as a flexible guide.

Early spring	March	Early autumn	September
Mid-spring	April	Mid-autumn	October
Late spring	May	Late autumn	November
Early summer	June	Early winter	December
Mid-summer	July	Mid-winter	January
Late summer	August	Late winter	February

Latin names have been given for all the plants in this book because these are the names that are universally used when describing plants; the Latin name should be recognised by the garden centre, and with any luck you will be sold the right plant. Common names have also been given, but these vary from country to country, and even within a country, and a plant may not be recognised by its common name.

Skill level is indicated by one of three ratings:
EASY, MEDIUM or **TRICKY**

Many of the plants chosen for this book have been given the Award of Garden Merit (AGM) by the Royal Horticultural Society (RHS). This is a really useful pointer in helping you decide which plants to buy. The AGM is intended to be of practical value to the ordinary gardener, and plants that merit the award are the cream of the crop. The RHS are continually assessing new plant cultivars and you can be sure that any plant with an AGM will have excellent decorative features and be:

- easily available to the buying public
- easy to grow and care for
- not particularly susceptible to pests or disease
- robust and healthy

SPRING

Spring is the most exciting time of the year for gardeners. All those bulbs you planted last year will be peeping through the soil now and patches of brown earth begin slowly receding, covered by emerging fresh, green plant growth. Don't be fooled that there will be plenty of rain – we are seeing drier springs and those early mini heat waves take us all by surprise. So why not invest in some spring-flowering plants or attractive foliage shrubs that have better drought resistance than some of the more traditional choices and will repay your interest in them a hundredfold by giving the best they've got, even when regular watering is limited.

Ajuga reptans
Bugle

⬆ 10cm/4in ⬌ 60cm/24in   EASY

This incredibly useful, low-maintenance, small, rhizomatous herbaceous perennial is native to Europe and Asia. It is especially charming for its evergreen, slightly quilted, dark green leaves, with the prettiest small spires of dark blue flowers in spring. It does well in humus-rich dry shade, but full sun seems to heighten the colour. The leaves may show signs of stress or slight crisping in bright sunlight and light soils (and it almost always gets mildew, so spray at the first sign of trouble).

BEST USES Delightful at the front of the border or edging paths, and incredibly effective for clothing awkward banks and slopes as ground cover; I have even known it survive under a large cedar

DROUGHT TOLERANCE Good, once established

FLOWERS April to May

SCENTED No

ASPECT West, north or east facing, in a sheltered position; full sun to partial shade

SOIL Any fertile, humus-rich, well-drained soil; add organic matter before planting

HARDINESS Fully hardy at temperatures down to -15°C/5°F; needs no winter protection

PROBLEMS Aphids; powdery mildew

CARE Cut back browning stems after flowering for a second flush of flowers and to keep it tidy, or run the mower over an unruly patch (with blades set high)

PROPAGATION Sow seed at 10°C/50°F in spring; division in spring or autumn; detach rooted plantlets and pot up in spring or early autumn

Aquilegia alpina
Columbine

⬆ 50cm/20in ⬌ 50cm/20in   **EASY**

Aquilegias are clump-forming perennials, originating in woodlands and meadows across the northern hemisphere, and are surprisingly drought tolerant for something so delicate looking, which, I dare say, is down to their thick root system. This variety is exceptionally pretty with bright green, very attractively lobed foliage and upright sprays of bright blue nodding flowers. Aquilegias are promiscuous and self-seed with abandon: you will soon have a thriving colony.

BEST USES An exuberant sight to behold when planted en masse, so use them as floral ground cover or mix them in amongst wildflowers and in herbaceous borders

DROUGHT TOLERANCE Good, once established
FLOWERS May to July
SCENTED No
ASPECT Any, in a sheltered or exposed position; full sun to partial shade
SOIL Any fertile, well-drained soil; add organic matter before planting
HARDINESS Fully hardy at temperatures down to -15°C/5°F; needs no winter protection
PROBLEMS Caterpillars and leaf miners; powdery mildew
CARE No cutting back is needed, but removing spent flowers will prevent too much self-seeding
PROPAGATION Self-seeds freely; sow ripe seed in a cold frame immediately or in spring (seed can take up to two years to germinate)

Aurinia saxatilis 'Citrina' ⚜
Alyssum saxatile

⬆ 20cm/8in ⬌ 30cm/12in   **EASY**

You may have to search around for this diminutive evergreen perennial from southern Europe, but it is certainly worth the effort. It has rosettes of grey-green leaves, forming attractive mounded clumps, and produces masses of slender stems, topped with a frenzy of pale lemon-yellow pincushion flower heads from late spring to summer.

BEST USES Super tumbling down dry banks, spilling over drystone walls or cascading in rock gardens; also easy to grow in pots or containers and provides excellent ground cover or border edging; very tolerant of coastal conditions, so an ideal choice for the seaside gardener

DROUGHT TOLERANCE Excellent, once established
FLOWERS May to June
SCENTED No
ASPECT South and west facing, in a sheltered position; full sun
SOIL Any fertile, well-drained soil; not suitable for heavy, wet soils
HARDINESS Fully hardy at temperatures down to -15°C/5°F; needs no winter protection
PROBLEMS Aphids
CARE Trim back after flowering for a second flush of flowers; cut back hard after final flowering to encourage new growth before winter
PROPAGATION Sow seed in a cold frame in autumn; softwood cuttings in early summer

Cerastium tomentosum
Snow-in-summer

⬆ 8cm/3in ⬌ Indefinite   EASY

This versatile, extremely vigorous mat-forming perennial from Sicily bears prolific small, tubular, flared, star-shaped white flowers against narrow, felty-soft silver-white leaves. Great value for money and much underrated.

> **BEST USES** Super tumbling down dry banks or in rock gardens, where it looks like a torrent of white water flowing downhill; thrives on dry banks or carpeting gravelled areas; also useful as border edging

DROUGHT TOLERANCE Excellent, once established

FLOWERS May to June

SCENTED No

ASPECT South and west facing, in a sheltered or exposed position; full sun

SOIL Any fertile, well-drained soil

HARDINESS Fully hardy at temperatures down to -15°C/5°F; needs no winter protection

PROBLEMS None

CARE Low maintenance, though may need cutting back to restrict its spread

PROPAGATION Sow seed in a cold frame in autumn; division in spring; stem-tip cuttings in late spring to early summer

··

GREENFINGER TIP *This can be invasive, so divide up unruly patches in early spring, before flowering begins*

Chamaemelum nobile
Wild/Roman chamomile

⬆ 30cm/12in ⬌ 45cm/18in   EASY

This popular diminutive, mat-forming perennial from Europe has slightly hairy filigreed foliage that releases a light apple scent when crushed underfoot and white, daisy-like flowers with yellow centres. It knits together fairly rapidly and is often used as a fragrant feature lawn. *C.n.* 'Treneague' is a less invasive, better-behaved aromatic variety: it has the same carpeting effect without flowers, making it a good choice for an ornamental lawn, but you may prefer to have the flowers.

> **BEST USES** Delightful as a quaint natural green path or lawn; perfect for creating a fragrant seat under a garden gazebo or herb bower

DROUGHT TOLERANCE Excellent, once established

FLOWERS May to July

SCENTED Aromatic leaves

ASPECT South, west or east facing, in a sheltered or exposed position; full sun

SOIL Any fertile, light, well-drained soil; will struggle in heavy, waterlogged soil

HARDINESS Fully hardy at temperatures down to -15°C/5°F; needs no winter protection

PROBLEMS None

CARE Trim the flower heads as they fade; when laid as a lawn, give a light haircut with a lawnmower to keep it compact

PROPAGATION Sow seed in situ in spring; division in spring

··

GREENFINGER TIP *It can be invasive on light soils*

Convolvulus cneorum ☻
Silverbush

⬆ 60cm/24in ↔ 75cm/30in   **EASY**

This enchanting evergreen shrub from the Mediterranean looks quite delicate but is really as tough as old boots. A member of the bindweed fraternity, but it does not have the twining, choking habit of common bindweed and is not invasive. It makes a small, rounded shrub of oval, silky, silver-grey leaves and pinkish, tightly furled, umbrella-like, coned flower buds that open up in true bindweed fashion to reveal an abundance of large, white, trumpet-shaped flowers with yellow centres.

BEST USES Great in rock gardens and alpine beds and very much at home in the Mediterranean or gravel garden; worthy of inclusion in a traditional herbaceous border, where its colouring provides a very effective foil for blue, purple and yellow flowering plants

DROUGHT TOLERANCE Excellent, once established

FLOWERS May to July

SCENTED No

ASPECT South or west facing, in a sheltered position; full sun

SOIL Any gritty, fertile, well-drained soil

HARDINESS Frost hardy at temperatures down to -5°C/23°F; needs winter protection in cold areas

PROBLEMS None

CARE Best left to its own devices, but a light trim after flowering in autumn will retain its bushy appearance and prevent it becoming straggly

PROPAGATION Softwood cuttings in late spring

Crepis incana ☻
Pink dandelion/Hawksbeard

⬆ 30cm/12in ↔ 30cm/12in   **EASY**

Don't be put off by the common name 'dandelion', which implies that this plant is yellow and very invasive. In fact, it is a very pretty, low-growing, free-flowering perennial, originally from Greece, from the aster family, with rayed, dandelion-like, ice-pink flowers, fine silver-grey foliage and familiar fluffy seed heads. It can be hard to come by nowadays, but is available at specialist nurseries and is well worth the hunt, as it is both pretty and unusual.

BEST USES Useful as an edging plant or planted in small drifts in a dry or gravel garden; looks equally pleasing in the cottage garden and attracts pollinating insects in fair abundance

DROUGHT TOLERANCE Excellent, once established

FLOWERS May to July

SCENTED No

ASPECT South or west facing, in a sheltered position; full sun

SOIL Any fertile, well-drained soil

HARDINESS Fully hardy at temperatures down to -15°C/5°F; needs no winter protection

PROBLEMS None

CARE None; best left to its own devices

PROPAGATION Self-seeds freely; sow ripe seed after flowering in a cold frame

Duchesnea indica
Mock strawberry

⬆ 10cm/4in ⬌ Indefinite   EASY

This semi-evergreen perennial from India, China and Japan looks just like a strawberry plant if you glance at it quickly. It has similar bright green, hairy leaves, small, pale yellow flowers and in late summer bears firm, bright red fruits that look very much like wild strawberries, but are bitter-tasting. It's a pretty, cheerful enough little thing, but incredibly invasive. (They did well in 6m/20ft of unimproved sandy soil in my own garden with only occasional rainfall, which demonstrates how out of hand they might get in ideal conditions.)

BEST USES Ideal as ground cover in a woodland garden or on a shady woodland bank, and will tolerate limited footfall; excellent at the front of a border and for the wildflower garden if you keep a close eye on its rampant inclination to spread

DROUGHT TOLERANCE Excellent, once established
FLOWERS March to May
SCENTED No
ASPECT Any, in a sheltered or exposed position; partial to full shade
SOIL Any fertile, humus-rich, well-drained soil; add organic matter before planting
HARDINESS Fully hardy at temperatures down to -15°C/5°F; needs no winter protection
PROBLEMS Slugs and snails may eat the fruits, but they are hard and sour so won't be missed
CARE Dig out unruly clumps to prevent spreading
PROPAGATION Sow seed in a cold frame in autumn; cut off the small rooted plantlets and pot them up

Erinus alpinus 🎖
Fairy foxglove/Jewel flower

⬆ 8cm/3in ⬌ 10cm/4in   EASY

This semi-evergreen perennial from central and southern Europe may be short-lived in plant terms, but it is well worth growing for its attractive, clump-forming habit, with small, mid-green, sticky leaves, and for the reliable profusion of dainty, pink to purple flowers with pale centres. An unbeatable flowerer, making pleasing clumps from spring to early summer.

BEST USES Very useful for growing in cracks in drystone walls, between paving slabs or in an alpine trough or container on the patio; makes an effective edging plant for borders or pathways

DROUGHT TOLERANCE Excellent, once established
FLOWERS May to June
SCENTED No
ASPECT Any, in a sheltered or exposed position; full sun to partial shade
SOIL Any fertile, well-drained soil; may struggle on heavy, wet clay
HARDINESS Fully hardy at temperatures down to -15°C/5°F; needs no winter protection
PROBLEMS None
CARE Trim lightly after flowering
PROPAGATION Self-seeds freely once established; sow seed in situ or in a cold frame in autumn; remove the new little side rosettes and pot up as cuttings in spring

Erodium × variabile 'Roseum' ☘
Stork's bill

⬆ 25cm/10in ⬌ 25cm/10in   **EASY**

Erodiums are members of the geranium family, and great little performers, but not as well known as they deserve to be considering their consistent flowering. From Europe and Central Asia, this perennial has a useful mat-forming habit, maturing to tidy cushions of small, lobed, mid-green leaves and a profusion of pale pink flowers with darker veining from late spring.

> **BEST USES** Useful for ground cover at the front or edge of borders and ideally suited to rock gardens and slopes; can also be grown successfully in containers or an alpine trough

DROUGHT TOLERANCE Excellent, once established
FLOWERS May to July
SCENTED No
ASPECT Any, in a sheltered position; full sun
SOIL Any fertile, humus-rich, gritty, well-drained soil; add organic matter before planting
HARDINESS Fully hardy at temperatures down to -15°C/5°F; needs no winter protection
PROBLEMS None
CARE None
PROPAGATION Sow ripe seed after flowering in a cold frame; division and basal cuttings in spring; if you are growing one or more varieties, seedlings may not come true, due to hybridisation

GREENFINGER TIP *Try to provide conditions that will spare them sitting in winter wet — they like sharp drainage all year*

EUPHORBIAS

Euphorbia griffithii 'Fireglow'

There are over 2000 species of euphorbia growing worldwide, from temperate and tropical zones in Asia, Europe and Africa. I have seen them growing wild on the poor, stony hillsides of southern Europe, dazzling the eye with their acid yellow flowers and needing no intervention by man to thrive in fairly impoverished conditions.

Owing to their widespread origins, there is a euphorbia for almost every situation in the garden. They can be annual, biennial or perennial shrubs and will tolerate a wide range of soils and pH levels, making them very versatile plants.

They are all easy to grow and low maintenance, though the milky sap exuded from their stems can be a skin irritant, so do wear gloves and, when taking cuttings, dip the cut ends into warm water to stem the bleeding sap. Apart from that one tiny flaw, every garden can benefit from planting a euphorbia: the only difficulty is deciding which of the many varieties to grow.

Many have attractive architectural form, whilst others are compact varieties that can cope happily in shade. Additional recommended varieties include *E. amygdaloides* var. *robbiae* with zingy lime flowers or the sulphuric yellow-flowered *E. polychroma* ☘ (both of which are compact).

Medium-sized *E. griffithii* 'Fireglow' ☘ has bold orange bracts lasting into summer; the smaller *E.* 'Redwings' offers bright yellow flowers from spring to summer and blue-green leaves, darkening to red in winter.

Euphorbia myrsinites ♐
Milk spurge

⬆ 10cm/4in ⬌ 30cm /12in   EASY

This unusually attired, prostrate evergreen perennial makes such an early, stealthy entrance that you can really luxuriate in its architectural development undistracted by its later-flowering competitors. With succulent, spiralled, diamond-shaped bluish grey-green leaves, it looks like a dozing snake uncoiling from a winter's sleep, looping its striking umbels of acid yellow flowers through the border. I saw this recently planted alongside black lilyturf (*Ophiopogon planiscapus* 'Nigrescens') and daffodils, and it looked spectacular. It must surely rate 9/10.

BEST USES Wonderful for the spring border and cottage garden; an obvious winner for the Mediterranean or gravel garden and on dry banks, and unusual but effective in containers

DROUGHT TOLERANCE Excellent, once established
FLOWERS May to June
SCENTED No
ASPECT South, east or west facing, in a sheltered or exposed position; full sun
SOIL Any fertile, well-drained soil
HARDINESS Fully hardy at temperatures down to -15°C/5°F; needs no winter protection
PROBLEMS None, though can suffer from aphids
CARE Cut flowering stems back after flowering
PROPAGATION Sow ripe seed in autumn; division after flowering; basal cuttings in spring or summer

Geranium macrorrhizum 'Ingwersen's Variety' ♐
Rock cranesbill

⬆ 50cm/20in ⬌ 50cm/20in   EASY

Originally from a range of temperate regions, geraniums will thrive in almost all conditions except the very wettest; this indispensable semi-evergreen hardy geranium will flourish where other herbaceous plants might petulantly sulk. It is a low-growing, mounded herbaceous perennial with pretty, pale pinky white flowers and attractive red stamens. The plum-scented, mid-green lobed leaves turn pale red in autumn.

BEST USES Perfect for ground cover in dry, shady or woodland areas; looks fabulous planted with spring bulbs, and the fresh green, mounded leaves help to disguise the bulbs' dying foliage

DROUGHT TOLERANCE Excellent, once established
FLOWERS May to July
SCENTED Lightly scented leaves
ASPECT Any, in a sheltered or exposed position; full sun to partial shade
SOIL Any fertile, well-drained soil; add organic matter before planting; will struggle in waterlogged clay
HARDINESS Fully hardy at temperatures down to -15°C/5°F; needs no winter protection
PROBLEMS Capsid bug and vine weevil; powdery mildew; slugs and snails
CARE Cut down to ground level in spring; cut back spent flowers to encourage further flowering
PROPAGATION Sow seed at 15°C/59°F in spring; division of large clumps in early spring or late summer or autumn; basal stem cuttings in spring

Hebe pinguifolia 'Pagei' ♀
Shrubby veronica

⬆ 30cm/12in ↔ 90cm/3ft   EASY

Hebes are local to mountain and coastal areas from Australia and New Zealand to South America. This variety is a compact, mat-forming evergreen shrub that boasts rounded, leathery, bluey green leaves which are a handsome feature all year round. Come late spring, it is smothered in masses of purple-stemmed white flowers. Needless to say, pollinating insects love it as much as we do.

BEST USES Impressive planted en masse as ground cover or as an undulating dwarf hedge; perfect for the city or courtyard garden in containers; will do well in coastal gardens

DROUGHT TOLERANCE Good, once established
FLOWERS May to June
SCENTED No
ASPECT Any, in a sheltered or exposed position; full sun to partial shade
SOIL Any reasonably fertile, well-drained soil
HARDINESS Fully hardy at temperatures down to -15°C/5°F; needs no winter protection
PROBLEMS Aphids; leaf spot and downy mildew
CARE Light trim in spring; little pruning needed
PROPAGATION Semi-ripe cuttings in a heated propagator in summer to autumn

GREENFINGER TIP *The centre of the plant may die off, so dig it up and plant small clumps in autumn to prevent it becoming all outer and no middle! The stems root as they grow: detach them with a clean, sharp penknife for an instant new plantlet*

Lamium maculatum
Spotted dead nettle

⬆ 20cm/8in ↔ 90cm/3ft   EASY

Lamium originate in Asia, Europe and north Africa, and are accustomed to anything from moist woodland to dry open spaces. This lovely ground-cover perennial has coarse green leaves with a definite silvery white variegation, which are an attractive feature in themselves. Small, pink, hooded flower spikes are borne in fair profusion in late spring and summer. It needs humus-enriched soil combined with shade to cope with drought, but nothing else offers as much cheerful daintiness in dry shade.

BEST USES Takes some beating as ground cover, forming a dense carpet quite rapidly and offering foliage that looks good nine months of the year

DROUGHT TOLERANCE Good, once established
FLOWERS May to July
SCENTED No
ASPECT West, north or east facing, in a sheltered position; full sun to deep shade
SOIL Any fertile, humus-rich, moist, well-drained soil; add organic matter before planting
HARDINESS Fully hardy at temperatures down to -15°C/50°F; needs no winter protection
PROBLEMS Slugs and snails can damage juvenile foliage
CARE Cut back after flowering for rapid new growth
PROPAGATION Sow seed in a cold frame in spring or autumn; division in spring

Papaver orientale 'Allegro'
Oriental poppy

⬆ 60cm/24in ⬌ 60cm/24in   EASY

This herbaceous, clump-forming perennial (originally from north Turkey and Iran) has hairy, toothed, greyish-green leaves and produces fat flower buds that burst open to reveal large, vibrant red-orange, papery petals with black-splashed centres. Ornamental, chubby salt 'n' pepper seed heads follow the flowers. Oriental poppies are plants of great, albeit fleeting, beauty and this bold colour makes a stunning visual impact.

BEST USES Glorious in a hot-colour border, making a striking contrast with grey foliage plants and deep purples; perfect for the cottage and wildlife garden as they are very attractive to bees

DROUGHT TOLERANCE Excellent, once established
FLOWERS May to July
SCENTED No
ASPECT South or west facing, in a sheltered position; full sun
SOIL Any deep, fertile, well-drained soil
HARDINESS Fully hardy at temperatures down to -15°C/5°F; needs no winter protection
PROBLEMS Aphids; downy mildew
CARE Stake plants; cut back after flowering if untidy; will die back naturally with first frosts
PROPAGATION Sow seed in situ in spring; division in spring; root cuttings in autumn

GREENFINGER TIP *Oriental poppies are pretty indestructible, but refuse to flower in overcast wet weather, so drought and plenty of sun suit them well*

Phlox divaricata 'May Breeze'
Wild sweet William

⬆ 30cm/12in ⬌ 20cm/8in   EASY

Phlox are an enormously diverse crowd, from as far afield as north America and Europe, so enjoy varying conditions from full sun to partial shade. Some are tall, for use in the border, some trailing, and others (like this one) are low-growing, mat-forming varieties. This evergreen, herbaceous perennial forms pleasing clumps of narrow mid-green leaves that look good all year. It bears a plethora of simple, lightly fragrant, gappy, five-petalled, white-tinted, ice-blue flowers in spring, which pollinating insects love.

BEST USES Ideal for traditional herbaceous borders, and a useful low-maintenance plant in the cottage or wildlife garden; equally at home in the gravel or Mediterranean garden

DROUGHT TOLERANCE Excellent, once established
FLOWERS April to June
SCENTED Lightly scented
ASPECT South, west or east facing in a sheltered or exposed position; full sun to partial shade
SOIL Any fertile, humus-rich, well-drained, gritty soil; add leafmould or organic matter before planting; will struggle in heavy clay
HARDINESS Fully hardy at temperatures down to -15°C/5°F; needs no winter protection
PROBLEMS Phlox eelworm; powdery mildew
CARE Deadhead after flowering
PROPAGATION Softwood cuttings in late spring

Phlox douglasii 'Boothman's Variety' ♛

⬆ 20cm/8in ↔ 30cm/12in   **EASY**

This modest little evergreen alpine perennial is grown for its foliage, which forms appealing evergreen mats of small, stiff, dark green leaves and bears pretty mauve flowers with violet centres. These are miniature plants in the strictest sense, so you really have to get your head down to appreciate their diminutive charms.

BEST USES A good choice for a rock garden, or a small raised garden border, or a trough planted alongside other rock plants to make a feature miniature garden. Also excellent in traditional borders as an edging plant and for both Mediterranean and gravel gardens

DROUGHT TOLERANCE Excellent, once established
FLOWERS April to June
SCENTED No
ASPECT South, west or east facing in a sheltered or exposed position; full sun to partial shade
SOIL Any poor, gritty, well-drained light soil; will struggle on heavy clay
HARDINESS Fully hardy at temperatures down to -15°C/5°F; needs no winter protection
PROBLEMS Phlox eelworm; powdery mildew
CARE Trim flower stems after flowering
PROPAGATION Softwood cuttings in late spring

Rosmarinus officinalis Prostratus Group
Trailing/Prostrate rosemary

⬆ 15cm/6in ↔ 1.5m/5ft   **EASY**

This low-growing evergreen shrub has small, narrow, very aromatic silver-grey foliage and a profusion of clear blue flowers in late spring to early summer. Not as well known as the common rosemary *R. officinalis* and nowhere near as hardy, it is nonetheless a delightful Mediterranean plant that adapts well to a variety of uses and deserves to be more popular.

BEST USES Fabulous as a low-growing, informal hedge; does well on dry banks and looks charming when left to spill carelessly over drystone walls; also ideal for coastal gardens

DROUGHT TOLERANCE Excellent, once established
FLOWERS May to June
SCENTED Aromatic leaves
ASPECT South or west facing, in a sheltered position; full sun
SOIL Any moderately fertile to poor, well-drained soil
HARDINESS Frost hardy at temperatures down to -5°C/23°F; needs winter protection in colder regions
PROBLEMS None
CARE Best left to its own devices; a light trim after flowering will stop it becoming leggy and woody
PROPAGATION Sow seed in a cold frame in spring; semi-ripe cuttings in summer to autumn; hardwood cuttings in late autumn to late winter

GREENFINGER TIP *This needs sharp drainage: it will struggle to do well in heavy clay soils*

Vinca minor f. *alba* 'Gertrude Jekyll'
Lesser periwinkle

⬆ 50cm/20in ⬌ 15cm/6in   EASY

There's no getting around it – I loathe periwinkle (largely because it is so often misused in unimaginative municipal planting schemes). Yet they are invaluable low-growing evergreen perennials, normally used as low-maintenance ground cover, and show remarkable drought tolerance if planted with organic matter. If I park my prejudice for a minute, this is a very attractive variety, with dark, glossy leaves and star-shaped, pure white flowers. *V. minor* 'Atropurpurea' has delicious damson-coloured flowers and dark green foliage.

BEST USES Wonderful ground cover, making a thick, weed-suppressing carpet in no time; useful for covering sloping banks, and 'Gertrude Jekyll' would look super in a shady woodland corner, with its dazzling white flowers

DROUGHT TOLERANCE Excellent, once established
FLOWERS April to September
SCENTED No
ASPECT Any, in a sheltered or exposed position; full sun to full shade
SOIL Any fertile, moist, well-drained soil; add organic matter before planting
HARDINESS Fully hardy at temperatures down to -15°C/5°F; needs no winter protection
PROBLEMS None
CARE Can become invasive, so trim straggling stems as they appear to prevent them rooting
PROPAGATION Division in spring or autumn; greenwood or semi-ripe cuttings at any time of year

Waldsteinia ternata

⬆ 10cm/4in ⬌ Indefinite   EASY

This seems to have lost favour of late, though it is an incredibly attractive and useful semi-evergreen, creeping, mat-forming perennial, originally from Europe, China and Japan. This variety has very pretty tri-lobed mid-green leaves and cheerful, buttercup yellow, shallow saucer-shaped flowers in spring.

BEST USES Definitely an ideal solution for that difficult spot in the garden where almost nothing else will grow; would suit a woodland garden, shady slope or inaccessible bank where maintenance is difficult

DROUGHT TOLERANCE Good, once established
FLOWERS April to June
SCENTED No
ASPECT Any, in a sheltered position; full sun to full shade
SOIL Any fertile, well-drained soil
HARDINESS Fully hardy at temperatures down to -15°C/5°F; needs no winter protection
PROBLEMS None
CARE Can become invasive, so divide unruly clumps after flowering and cut back at intervals to curb its spread
PROPAGATION Sow ripe seed in a cold frame in autumn; division in early spring

Danae racemosa
Alexandrian laurel/Poets laurel

⬆ 90cm/3ft ↔ 90cm/3ft   MEDIUM

This slightly unusual, evergreen, shrubby perennial from Turkey is infrequently seen nowadays, yet it has much to commend it, with handsome, pointed, highly polished dark green leaves and an upright habit with arching green stems. Greenish flowers are followed by orangey red berries, making it ideal for offering year-round garden interest. An excellent low-maintenance plant that deserves to be more popular.

BEST USES Useful for borders and as ground cover; will perform equally well as an evergreen hedge as it takes to clipping admirably

DROUGHT TOLERANCE Excellent, once established
FLOWERS June, but grown mainly for foliage and autumn berry display
SCENTED No
ASPECT Any, in a sheltered position; partial to full shade
SOIL Any fertile, well-drained soil; add organic matter before planting
HARDINESS Fully hardy at temperatures down to -15°C/5°F; needs no winter protection
PROBLEMS None
CARE Prune dead or damaged stems back to ground level in spring
PROPAGATION Sow seed in a cold frame in autumn; division in early spring or early autumn

Elymus hispidus
Blue wheat grass

⬆ 75cm/30in ↔ 40cm/16in   EASY

This has got to be one of the steeliest blue semi-evergreen grasses around, with its stunning, silvery blue, slightly arching linear leaves, bearing rather insignificant parchment-coloured panicles in summer. However, it is not the flowers you are after, but the attractive clump-forming foliage, whose leaves seem to colour up in full sun. It can be invasive, so keep an eye on it.

BEST USES Ideal studding the ground in the gravel or dry garden; also looks good in containers in a contemporary city garden

DROUGHT TOLERANCE Excellent, once established
FLOWERS July, but grown mainly for foliage
SCENTED No
ASPECT South or west facing, in a sheltered position; full sun
SOIL Any fertile, well-drained soil
HARDINESS Fully hardy at temperatures down to -15°C/5°F; needs no winter protection
PROBLEMS Rust
CARE Cut to ground level in winter to remove dead leaves
PROPAGATION Sow seed in situ in spring; division in spring

GREENFINGER TIP *Grow in pots to curb its invasive tendencies, if you are worried about it wandering*

Euphorbia characias 'Black Pearl'

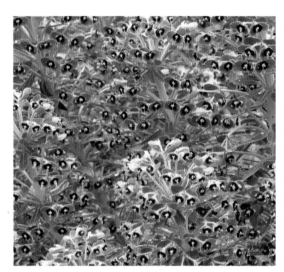

↑ 75cm/30in ⬌ 75cm/30in   EASY

This evergreen spurge has a striking, architectural form with stiff, upright stems and whorls of narrow grey-green leaves with upright, clustered, zingy lime green flowers that are so bright as to be dazzling. Each cupped flower has an appealing browny-black eye. This is an eye-catching low-maintenance plant, ideal for foliage lovers. You may have to search around for it, but it will be worth it.

> **BEST USES** Ideal in the Mediterranean and gravel garden as well as traditional borders; also looks at home in a chic city garden or small sunny patio

DROUGHT TOLERANCE Excellent, once established

FLOWERS May to July, but also grown for foliage and architectural form

SCENTED No

ASPECT South, west or east facing, in a sheltered position; full sun

SOIL Any fertile, well-drained soil

HARDINESS Fully hardy at temperatures down to -15°C/5°F; needs no winter protection

PROBLEMS None

CARE Cut down flowering stems to ground level after flowering

PROPAGATION Sow ripe seed in a cold frame in autumn (germination can be erratic); division in spring; basal cuttings in early summer

HARDY GERANIUMS

Geranium x *riversleaianum* 'Russell Prichard'

Never underestimate the value of the more common herbaceous perennials: the hardy geraniums are indispensable in the gardening year. (They are often confused with half-hardy or greenhouse geraniums, which are not winter hardy and are known as pelargoniums.)

There are over 400 hardy geraniums, also known as cranesbill because of their beak-like seed casings. Originally from the mountain areas of France, the Balkans and across to western Russia, they succeed in wide and varying habitats, offering great decorative interest for comparatively little effort. Very easy to grow, they will put up with almost any soil so long as it is well drained. They are generally disease free and require little maintenance.

All hardy geraniums will grow in full sun to partial shade and they are effectively drought resist-ant once established, if planted in humus-rich soil from the outset (though of course some varieties do better than others). The leaves may crisp in times of prolonged drought, but most will survive.

They are invaluable for dry gardens, and also for the herbaceous border or woodland garden: there is a plant to suit just about every garden aspect, with a great many wonderful varieties to choose from in addition to those listed, but the ones included here will do well in drier conditions. If you don't have any hardy geraniums in your garden, now is the time to invest in some for their ever-willing capacity to thrive on paucity.

Geranium phaeum
Mourning widow geranium

⬆ 80cm/32in ⬌ 45cm/18in   EASY

A tall herbaceous perennial with divided mid-green leaves and small, reflexed, dusky chocolate-maroon flowers. This is a particularly attractive plant that does better in dry shade than any of the other hardy geraniums if improved with humus first, though the leaves can scorch in prolonged drought.

BEST USES Does well in a light woodland garden; clumps up over time so useful as ground cover on shallow slopes; also marvellous underplanted in a shrubbery and happily earns its keep in a traditional flower border

DROUGHT TOLERANCE Excellent, once established

FLOWERS May to July

SCENTED No

ASPECT Any, in a sheltered or exposed position; full sun to partial shade

SOIL Almost any well-drained soil; add organic matter before planting; will struggle in boggy or waterlogged soil

HARDINESS Fully hardy at temperatures down to -15°C/5°F; needs no winter protection

PROBLEMS None

CARE Cut back to ground level in early spring to promote new growth; remove old flowering stems and faded leaves in summer after flowering

PROPAGATION Sow seed at 15°C/59°F in spring; division in spring

Libertia grandiflora 🎖
New Zealand satin flower

⬆ 90cm/3ft ⬌ 90cm/3ft   EASY

This rhizomatous evergreen perennial from New Zealand makes handsome rounded clumps of leathery, sword-like, bright green leaves and has tall flower spikes of small, pure white flowers that are a little iris-like in appearance, followed by orange seed heads that split to reveal appealing shiny black seeds. It is not as popular as might be expected, especially as the seed heads offer prolonged interest after fairly long-lasting flowers. I discovered it at Beth Chatto's garden and if it's good enough for her, it's certainly good enough for us.

BEST USES Ideal for prairie-style gardens, and an elegant addition to any flower border, but looks equally well in a cottage border or gravel garden

DROUGHT TOLERANCE Excellent, once established

FLOWERS May to July

SCENTED No

ASPECT South or west facing, in a sheltered position; full sun

SOIL Any fertile, well-drained soil; will struggle in heavy clay

HARDINESS Fully hardy/borderline at temperatures down to -15°C/5°F; may need winter protection in colder areas

PROBLEMS None

CARE Cut back the flower spikes after flowering

PROPAGATION Self-seeds easily; sow ripe seed (extracted from the berries) at 10°C/50°F; division in spring

Milium effusum 'Aureum' ♈
Bowles' golden grass

⬆ 90cm/3ft ⬌ 30cm/12in   EASY

A spreading semi-evergreen perennial from Europe and the grasslands of North America, this is another plant that can be endowed with good drought tolerance if planted in humus-rich soil from the outset. It makes clumps of flat-bladed, achingly bright golden, slender graceful leaves in spring, when the foliage is at its brightest, followed by airy panicles of tiny golden flowers.

> **BEST USES** Ideal in dappled woodland or a shady corner where the bright gold foliage lifts the gloom and makes effective ground cover; naturalises beautifully with aquilegia; great for the urban garden as it is both modern and low maintenance

DROUGHT TOLERANCE Good, once established

FLOWERS May to July

SCENTED No

ASPECT Any, in a sheltered or exposed position; partial shade

SOIL Any fertile, well-drained soil; add organic matter before planting

HARDINESS Fully hardy at temperatures down to -15°C/5°F; needs no winter protection

PROBLEMS None

CARE Cut down to ground level in late autumn or early spring

PROPAGATION Self-seeds easily; sow seed in situ in spring; division in spring or early summer

GREENFINGER TIP *If the leaves start looking tatty, just give it a haircut to stimulate fresh growth*

Salvia officinalis 'Purpurascens' ♈
Purple sage

⬆ 80cm/32in ⬌ 90m/3ft   MEDIUM

Common perhaps, but this is an absolutely indispensable evergreen sub-shrub from the Mediterranean and north Africa, providing year-round interest. The aromatic, soft, felty, pointed, oval sage green leaves are suffused with purple and make a pleasing mound with very pretty branching spires of soft lilac flowers in summer. Traditionally it has great healing properties.

> **BEST USES** Invaluable in the herb garden, but equally at home in the cottage garden, where it combines well with gold-leaved foliage or yellow or deep-coloured perennials; ideal in a Mediterranean garden, and a great addition to the wildflower garden as bees love it too

DROUGHT TOLERANCE Excellent, once established

FLOWERS May to July

SCENTED Aromatic leaves

ASPECT South, west or east facing, in a sheltered or exposed position; full sun to partial shade

SOIL Any fertile, well-drained soil

HARDINESS Fully hardy at temperatures down to -15°C/5°F; needs no winter protection

PROBLEMS Can be temperamental in establishing; slugs and snails like to nibble the young foliage

CARE Prune very lightly in summer after flowering

PROPAGATION Softwood cuttings in late spring or early summer

GREENFINGER TIP *The trick to growing this successfully is to keep the hacking to a minimum!*

Acanthus spinosus ♉
Bear's breeches

⬆ 1.5m/5ft ⬌ 60cm/24in   EASY

An architectural semi-evergreen perennial from the Mediterranean that has large, glossy, lobed leaves up to 60cm/24in long and the most marvellous, spiny, plum-coloured flower spikes, with white-pink lips. This plant is a great performer, looking pretty good for at least nine months of the year, though the roots can be fairly invasive. I grew this in my garden in Spain, so I can fully vouch for its drought pedigree. *A. mollis* is also worth considering; the flowers are very similar to *A. spinosus*, but it has rounder leaves.

BEST USES Display as a long-lasting sculptural plant in a spacious flower border; a tremendous woodland plant, and attractive for wildflower havens as it is loved by pollinating insects

DROUGHT TOLERANCE Excellent, once established
FLOWERS May to July
SCENTED No
ASPECT Any, in a sheltered or exposed position; full sun to partial shade
SOIL Any fertile, well-drained soil
HARDINESS Fully hardy at temperatures down to -15°C/5°F; needs no winter protection
PROBLEMS None
CARE Cut down to ground level in late winter or early spring; divide overcrowded clumps every four years in early spring or autumn
PROPAGATION Sow seed in a cold frame in spring; division in spring or autumn; root cuttings in winter (they may take up to two years to flower)

Ceanothus 'Blue Mound' ♉
Californian lilac 'Blue Mound'

⬆ 1.5m/5ft ⬌ 2m/6ft   EASY

The Californian lilacs are a really useful group of low-maintenance shrubs with profusions of (usually) blue-hued flowers. They are reliable, problem free and beautiful in flower, and also adapt well as climbing wall shrubs. This mound-forming evergreen variety is fairly vigorous, with glossy, delicately toothed dark green leaves and pure dark blue flowers, borne in quantity in spring to early summer. The sunnier the position, the more profuse the flowering. C. 'Puget Blue' ♉ is a much larger, deep blue variety.

BEST USES Useful at the back of a Mediterranean border, providing all-year interest; very effective as low-maintenance ground cover on a hot sunny bank or awkward slope; bees love it

DROUGHT TOLERANCE Excellent, once established
FLOWERS April to June
SCENTED No
ASPECT South, west or east facing, in a sheltered position protected from cold winds; full sun
SOIL Any fertile, well-drained soil
HARDINESS Frost hardy at temperatures down to -5°C /23°F; needs winter protection in all but the mildest areas
PROBLEMS None
CARE Trim lightly after flowering to maintain spread and size
PROPAGATION Semi-ripe cuttings in summer to autumn; hardwood cuttings in autumn to late winter

Choisya ternata ♛
Mexican orange blossom

⬆ 2.5m/8ft ⬌ 2.5m/8ft   **EASY**

A delightful compact evergreen shrub from Mexico, very widely grown, but none the worse for its familiarity as it really is a great asset to the garden, providing exuberant year-round interest. It has glossy, dark green, oval leaves, with clusters of small, white, sweetly fragrant flowers in spring and flowers again in late summer.

BEST USES Perfect for a hot, sunny mixed border or to add an exotic touch to a shady border; very effective as an informal flowering hedge; useful for city gardens as it is tolerant of pollution

DROUGHT TOLERANCE Excellent, once established

FLOWERS May, with repeat flowering in summer and autumn

SCENTED Lightly perfumed flowers

ASPECT South, west or east facing, in a sheltered position; full sun to partial shade

SOIL Any fertile, well-drained soil; add organic matter before planting

HARDINESS Fully hardy at temperatures down to -15°C/5°F; needs no winter protection

PROBLEMS Snails may nibble young plants

CARE Trim lightly after flowering to maintain spread and size; takes hard pruning well if gets big for its space

PROPAGATION Semi-ripe cuttings in mid-summer to autumn

GREENFINGER TIP *People often plant this in shady spots because it does okay there, but flowering is reduced significantly; flower and leaf are both superior in full sun*

Cytisus × *praecox* 'Warminster' ♛
Broom

⬆ 1.5m/5ft ⬌ 1.5m/5ft   **EASY**

This is a very dependable, though often short-lived, small, bushy, rounded deciduous shrub with green arching stems which bear a cheerful profusion of pea-like, pungent, creamy yellow flowers in spring before the small, grey-green leaves appear. C. 'Boskoop Ruby' ♛ is a rather more unusual, rich red-flowered, less hardy variety, and well worth considering if you are not a great lover of yellows.

BEST USES Will look good in any mixed shrub border; early pollinating insects love it, so a good choice for the cottage or wildflower garden; will do well in coastal areas

DROUGHT TOLERANCE Excellent, once established

FLOWERS April to May

SCENTED Yes

ASPECT South, west or east facing, in a sheltered or exposed position; full sun

SOIL Any fertile, well-drained soil

HARDINESS Fully hardy at temperatures down to -15°C/5°F; needs no winter protection

PROBLEMS None

CARE Trim lightly after flowering to maintain its rounded habit

PROPAGATION Semi-ripe cuttings in summer; hardwood cuttings in mid-winter

Euphorbia characias subsp. *wulfenii* ♀
Mediterranean spurge

⬆ 1.2m/4ft ⬌ 1.2m/4ft   EASY

This evergreen spurge from the Mediterranean is a real winner. It has much to offer in year-round interest, with tall erect stems that are whorled with sage green lance-shaped leaves. Very attractive, striking, lime green flower heads or umbels, which are architectural in shape, are produced in spring. I am very fond of this plant, though earlier in my career I got quite good at killing it because I cut it back at the wrong time of year! You live and learn.

BEST USES Adds excellent 'bones' to the herbaceous border, Mediterranean or gravelled area, and is equally at home in a contemporary or traditional garden: 9/10

DROUGHT TOLERANCE Excellent, once established

FLOWERS March to May

SCENTED No

ASPECT South, west or east facing, in a sheltered or exposed position; full sun

SOIL Any fertile, well-drained soil

HARDINESS Fully hardy at temperatures down to -15°C/5°F; needs no winter protection

PROBLEMS None, though botrytis (grey mould) can, rarely, be a nuisance

CARE Remove the spent flowering stems after flowering, as close to the base as you can

PROPAGATION Sow ripe seed in a cold frame in autumn (germination can be erratic); basal cuttings in early summer

Lupinus arboreus ♀
Tree lupin

⬆ 1.5m/5ft ⬌ 1.5m/5ft   EASY

If you have the room for fleeting fancy, this Californian native is a slightly sprawling semi-evergreen shrub (unlike the herbaceous lupins) with very delicate filigreed, grey-green foliage that shows off the sweetly scented, candled spires of lemon, pea-like flowers, up to 30cm/12in long, to great advantage in late spring to summer.

BEST USES I grew this dotted through a wildflower meadow and it looked heavenly, but it will look quite at home in coastal and cottage gardens alike

DROUGHT TOLERANCE Excellent, once established

FLOWERS May to June

SCENTED Yes

ASPECT South or west facing, in a sheltered position; full sun

SOIL Any fertile, well-drained soil, particularly sandy or loam

HARDINESS Fully hardy at temperatures down to -15°C/5°F; needs no winter protection

PROBLEMS Aphids (including lupin greyfly), slugs and snails; powdery mildew

CARE Remove spent flower spikes

PROPAGATION Softwood cuttings in spring

....

GREENFINGER TIP *Very unusually, it can send out a bluish flower: just remove it as soon as you spot it*

Rosa banksiae 'Lutea'
Yellow banksian rose

⬆ 9m/30ft ⬌ 9m/30ft   EASY

This is enduringly popular because it's an early flowering rose, is evergreen (though only in the mildest areas) and almost thornless – but don't be fooled, the odd thorn will take you unawares! Originally from China, it has small, lightly fragrant clusters of pale primrose-yellow flowers (2cm/¾in across) that burst forth in fair abundance from late spring to early summer.

BEST USES Delightful for training along a fence and a splendid sight against a large, warm sheltered wall; combines incredibly well with wisteria, which flowers at much the same time

DROUGHT TOLERANCE Good, once established

FLOWERS May to June

SCENTED Lightly fragrant flowers

ASPECT South, west or east facing, in a sheltered position; full sun

SOIL Any fertile, humus-rich, moist, well-drained soil; add organic matter before planting

HARDINESS Frost hardy at temperatures down to -5°C/23°F; needs winter protection

PROBLEMS Aphids and caterpillars; blackspot, powdery mildew and rust

CARE Mulch the base with organic matter in spring; deadhead after flowering, and cut out dead or damaged material in late autumn and early spring; no initial pruning, but every three years cut back three of the main stems to 45cm/18in above the ground to maintain shape, health and vigour

PROPAGATION Hardwood cuttings in autumn

Rosa Summer Song
(also known as 'Austango')

⬆ 1.2m/4ft ⬌ 90cm/3ft   EASY

Some roses need a lot of nursing, but there are modern cultivars that are not only disease resistant but also do well in dry weather. Planted with humus-rich soil and given annual organic mulches, this makes a very dependable drought-tolerant shrub. R. Summer Song is a bushy, upright, deciduous English shrub rose with glossy, dark green leaves and sumptuous rich, some say, burnt orange, unfading, double-cupped blooms with a distinct fruity fragrance.

BEST USES Perfect for the mixed border with purples or soft yellows and apricots; can be grown in a container by pruning to restrict size

DROUGHT TOLERANCE Excellent, once established

FLOWERS May to June

SCENTED Yes

ASPECT South or west facing, in a sheltered position; full sun

SOIL Any fertile, well-drained soil; add organic matter before planting

HARDINESS Fully hardy at temperatures down to -15°C/5°F; needs no winter protection

PROBLEMS Greenfly; has good disease resistance

CARE Mulch annually with organic matter; deadhead after flowering and cut back flowered stems, leaving 2–3 buds on this season's new growth; prune back by a third in winter or when growth starts in spring in colder regions

PROPAGATION Hardwood cuttings in autumn

Rosmarinus officinalis
Rosemary

⬆ 1.5m/5ft ⬌ 1.5m/5ft   EASY

This familiar and much-loved evergreen shrub has graced many a roast leg of lamb and is much sought after for its versatility in the kitchen. In plant terms, it is an attractive, bushy, aromatic shrub, with felty, needle-like, silvery-grey leaves and produces an abundance of small, pale blue-purple flowers in spring to summer, but often flowering earlier than expected. Bees love it!

BEST USES Ideal for a hot, sunny border or dry bank, and there's always a spot for this outside the kitchen door, so you can grab handfuls of the scented foliage for your culinary exploits

DROUGHT TOLERANCE Excellent, once established
FLOWERS May to June
SCENTED Aromatic leaves
ASPECT South or west facing, in a sheltered position; full sun
SOIL Any fertile, well-drained soil
HARDINESS Frost hardy at temperatures down to -5°C/23°F; needs winter protection
PROBLEMS Aphids
CARE Trim lightly after flowering to prevent the plant becoming leggy and to retain its bushy habit
PROPAGATION Sow seed in a cold frame in spring; semi-ripe cuttings in summer to autumn; hardwood cuttings in late autumn to late winter

GREENFINGER TIP *Propagation by cuttings will yield good-sized plants more quickly than sowing seed*

Tamarix tetrandra ♈
Tamarisk

⬆ 4m/13ft ⬌ 3m/10ft   EASY

A compact, gracefully arching, deciduous shrub or small tree that bears attractive smoky wisps of pale pink flowers in spring. The stems are a striking mulberry colour, the needled leaves a fresh, pleasing green. It is a low-maintenance plant and, despite looking so fragile, is tough as old boots which, given its origins in Europe, Asia and India, should come as no surprise.

BEST USES Particularly good for coastal gardens as it is tolerant of salty air, so invaluable as a coastal windbreak or informal hedge

DROUGHT TOLERANCE Excellent, once established
FLOWERS April to May
SCENTED No
ASPECT Any, in a sheltered or exposed position; full sun
SOIL Any fertile, well-drained soil; will not tolerate waterlogged soils
HARDINESS Fully hardy at temperatures down to -15°C/5°F; needs no winter protection
PROBLEMS None
CARE Trim lightly after flowering
PROPAGATION Sow ripe seed in a cold frame in spring; semi-ripe cuttings in summer; hardwood cuttings in winter

GREENFINGER TIP *After a few years the plant can get slightly overbalanced and top heavy, in which case just cut back the offending flowering stems or branches in spring*

Ulex europeaus
Common gorse

⬆ 2.5m/8ft ⬌ 2m/6ft   **EASY**

This very dense, evergreen shrub, native to western Europe, has narrow, deep green leaves. Its stems are covered with thorny spines, but this is more than compensated for by the coconut-scented, gaudy yellow flowers that smother it in spring. It is regarded as a common shrub in the wild, but is as tough as they come: if it had a snottier reputation we would all find space for it.

> **BEST USES** Ideal in coastal gardens as hedging or windbreaks (and the thorny spines deter animal intruders); pollinating insects swarm all over it

DROUGHT TOLERANCE Excellent, once established

FLOWERS April to June

SCENTED Scented flowers

ASPECT Any, in a sheltered or exposed position; full sun

SOIL Any fertile, well-drained soil

HARDINESS Fully hardy at temperatures down to -15°C/5°F; needs no winter protection

PROBLEMS None

CARE Cut out faded flower spikes and old dead growth in spring

PROPAGATION Pre-soak seed and sow in spring or autumn; semi-ripe cuttings in summer; hardwood cuttings in late autumn to mid-winter

Wisteria sinensis 🎖
Chinese wisteria

⬆ 15m/50ft ⬌ 12m/40ft   **EASY**

This vigorous deciduous climber has long, pendulous, sweetly perfumed, pale lilac-blue flowers, resembling a splendid tiered chandelier. They are produced in abundance on bare stems, with the light green pinnate leaves appearing when the flowers fade, and are followed by long velvety pea pods. Always buy wisteria in flower, as initial flowering can take up to seven years.

> **BEST USES** Leave to scramble through a large tree or shrub; grow on a strong pergola in a Mediterranean garden, or as a standard tree

DROUGHT TOLERANCE Excellent, once established

FLOWERS May to June

SCENTED Highly perfumed flowers

ASPECT South or west facing, in a sheltered position; full sun (though will flower reasonably well in partial shade)

SOIL Any fertile, moist, well-drained soil; add organic matter before planting

HARDINESS Fully hardy at temperatures down to -15°C/5°F; needs no winter protection

PROBLEMS Frosts may damage flower buds, particularly on east-facing walls; powdery mildew

CARE Tie in stems in August and cut back this season's shoots to about 30cm/12in from the point they have grown from; in February cut these shoots back again, to within 10–12cm/4–5in (2–3 buds) of the old wood; cut the long, whippy shoots that grew after the summer to about 5 or 6 buds from the main branch

PROPAGATION Layering in spring

SUMMER

Drought-tolerant plants really come into their own during the long, dry summer months. The proof of the pudding is in the eating, they say – so if you have been wise enough to select some of the more drought-resistant plants available, you can take time for a small moment of self-congratulation, as you won't be darting madly about with the watering can, agonising helplessly when your promising young plants shrivel to grey ash in the midday sun. Introducing even two new drought-tolerant plants each year will lead to the garden becoming more lovely and needing less maintenance each season.

Alchemilla mollis 🎖
Lady's mantle

⬆ 60cm/24in ⬌ 60cm/24in   **EASY**

Some plants can't be praised too highly, and this tough herbaceous perennial from the mountains and meadows of Turkey is one of them. It forms low mounds of softly scalloped limey leaves that look stunning when jewelled with raindrops. Delicate sprays of acid yellow flowers are held erect on long stems in summer, lasting for weeks. This good-natured plant needs so little care that it must rate 9/10 for consistent performance. In extreme drought, the edges of the leaves may crisp and brown a little but, frankly, you'll have to overlook this one tiny flaw.

BEST USES Delightful at the front of the border, edging paths or as a swathe of ground cover (I have even known it thrive under an Atlas cedar)

DROUGHT TOLERANCE Good, once established
FLOWERS June to September
SCENTED No
ASPECT Any, in a sheltered or exposed position; full sun to partial shade
SOIL Any fertile, well-drained soil
HARDINESS Fully hardy at temperatures down to -15°C/5°F; needs no winter protection
PROBLEMS None
CARE Cut the whole plant to ground level in spring; cut back browning flower heads to encourage a second flush of flowers and to prevent invasive self-seeding
PROPAGATION Self-seeds freely; sow seed in a cold frame in spring; division in spring or autumn

Allium cristophii
Star of Persia

⬆ 60cm/24in ⬌ 20cm/8in   **EASY**

It is easy to see why this startling sputnik of a plant has captured the hearts of gardeners and designers alike. Originally from Turkey, this bulbous perennial has strappy, fresh green leaves like many of the large alliums, but it's the huge (20cm/8in), globular, purple flower heads, made up from myriad tiny, star-shaped flowers and borne singly on bare stems, that set this apart as an object of great wonder. The leaves will suffer from browning in prolonged drought.

> **BEST USES** Fantastic as a vertical accent in the herbaceous border; bees and butterflies adore it, so ideal for encouraging pollinating insects

DROUGHT TOLERANCE Good, once established

FLOWERS June to July

SCENTED Onion-scented leaves when crushed

ASPECT South, west or east facing, in a sheltered or exposed position; full sun

SOIL Any fertile, well-drained soil; add organic matter before planting

HARDINESS Frost hardy at temperatures down to -5°C/23°F; needs winter protection in colder areas

PROBLEMS Onion fly; onion white rot and powdery mildew

CARE Foliage dies down naturally with the onset of winter but the flower heads take on an attractive skeletal form, so don't cut them down

PROPAGATION Sow ripe seed in a cold frame in spring; division in late summer

Anaphalis margaritacea
Pearly everlasting

⬆ 60cm/24in ⬌ 60cm/24in   **EASY**

This is a bushy, clump-forming rhizomatous perennial from Asia and North America. It has an upright habit, with woolly, silver-grey leaves with white felty undersides and bears clusters of small, rounded pearl-white flowers with yellow eyes in summer; the flowers turn to a grubby pale yellow. Because of its compact habit, large swathes make handy ground cover in dry areas. It does a job, but it's not one of my favourites.

> **BEST USES** Useful as low level ground cover or near the front of borders; especially effective in a white border, with its woolly masses of silver-white flowers and muted foliage; the cut flowers appeal to flower arrangers as they last well

DROUGHT TOLERANCE Excellent, once established

FLOWERS June to September

SCENTED No

ASPECT Any, in a sheltered or exposed position; full sun to partial shade

SOIL Any fertile, moist, well-drained soil; add organic matter before planting

HARDINESS Fully hardy at temperatures down to -15°C/5°F; needs no winter protection

PROBLEMS None

CARE Cut back hard in spring to prevent it becoming straggly

PROPAGATION Sow seed in a cold frame in spring; division in spring; basal stem cuttings in spring

Anthemis punctata subsp. *cupaniana* ♀
Sicilian chamomile

⬆ 30cm/12in ⬌ 90cm/3ft   **EASY**

This tough evergreen perennial from Sicily has aromatic, fringed, silver-grey leaves that green up in winter and is reasonably fast-growing, forming fairly dense matting quite quickly. It bears a profusion of pretty, cheerful, white daisy-like flowers with bright yellow centres 6cm/2in across for weeks on end from early summer. A dependable, long-flowering plant that is useful for crevices where nothing else will grow.

BEST USES Great as low-level ground cover or at the front of borders; marvellous for clothing hot, dry sunny banks or rockeries; attractive to bees and other pollinating insects, so ideal for the wildflower or cottage garden

DROUGHT TOLERANCE Good, once established

FLOWERS May to August

SCENTED No

ASPECT South, west or east facing, in a sheltered or exposed position; full sun

SOIL Any fertile, well-drained soil

HARDINESS Frost hardy at temperatures down to -15°C/5°F; needs winter protection in colder areas

PROBLEMS Slugs; powdery mildew

CARE Cut back faded flower stems to encourage new growth

PROPAGATION Sow seed in a cold frame in spring; division in spring

Armeria maritima 'Vindictive' ♀
Thrift/Sea pink

⬆ 50cm/20in ⬌ 50cm/20in   **EASY**

This thrift comes originally from northern hemisphere coastal areas, so it is tough and adaptable. A useful, mat-forming evergreen perennial often grown as an alpine or rock plant, it has fine-needled green leaves and forms neat hummocks from which appear small clusters of rounded, dark rosy-pink flowers in summer.

BEST USES Ideal for rock and gravel gardens and as edging at the front of borders, or for growing in a trough on a sunny patio with other alpine plants; very suitable for coastal conditions as it is tolerant of salt-laden winds

DROUGHT TOLERANCE Excellent, once established

FLOWERS May to September

SCENTED No

ASPECT Any, in a sheltered or exposed position; full sun

SOIL Any fertile, well-drained soil; will struggle on heavy, wet clay

HARDINESS Fully hardy at temperatures down to -15°C/5°F; needs no winter protection

PROBLEMS None

CARE No cutting back is needed, but remove spent flowers, if desired

PROPAGATION Sow seed in a cold frame in early spring or early autumn; division in early spring or autumn

Artemisia 'Powis Castle' 🎖
Wormwood/Mugwort

⬆ 60cm/24in ⬌ 90cm/3ft   **EASY**

Artemisia 'Powis Castle' has the finest foliage of all the artemisias, with its feathery, finely cut, strongly aromatic silver-grey leaves. This woody perennial from the eastern Mediterranean rarely bears flowers, but when they do appear they are small, insignificant and yellow-white. I tend to shear them off, because I don't like them, leaving an excellent foliage plant. Artemisias can look a little straggly in winter, but as drought lovers with outstanding foliage they take some beating.

BEST USES Perfect in a hot, dry border in a gravel or Mediterranean garden; the foliage also suits the cottage garden (I grow them at the front of the border as a swathe of very appealing edging)

DROUGHT TOLERANCE Excellent, once established
FLOWERS August, though grown mainly for foliage
SCENTED Aromatic leaves
ASPECT South, west or east facing, in a sheltered or exposed position; full sun
SOIL Any fertile, well-drained soil; hates winter wet and will be short-lived on heavy, poor-draining soils
HARDINESS Frost hardy at temperatures down to -5°C/23°F; needs winter protection in colder areas
PROBLEMS Aphids
CARE Trim lightly after flowering and cut back to the base in autumn
PROPAGATION Division in spring; heeled greenwood cuttings in mid- to late summer (take regular greenwood cuttings as this plant can be shortlived)

Ballota pseudodictamnus 🎖
False dittany

⬆ 50cm/20in ⬌ 60cm/24in   **EASY**

This woody-based, evergreen, low-growing sub-shrub, originally from the Mediterranean, has neat rosettes of felty, grey-green rounded leaves that make a low hummock. The knobbly bobbles along the stems open to tiny, pinkish-white flowers, but it is more popularly grown for its pretty foliage, as the flowers are unremarkable.

BEST USES Perfect for border edging or ground cover, and very attractive planted in multiples as a staggered row of woolly hummocks; useful for city gardens, in containers and hanging baskets

DROUGHT TOLERANCE Excellent, once established
FLOWERS August to September, but grown mainly for foliage
SCENTED No
ASPECT South or west facing, in a sheltered or exposed position; full sun
SOIL Any fertile, well-drained soil; may struggle on heavy, wet clay
HARDINESS Fully hardy/borderline at temperatures down to -15°C/5°F; may need winter protection in colder areas
PROBLEMS None, though may loll in rain as the felty leaves act like sponges, but soon dries out
CARE Cut back in late spring before new growth starts, to stop it getting leggy (cutting back every other year may be sufficient, but the central stems will get woollier in the second year)
PROPAGATION Softwood cuttings in late spring to early summer; semi-ripe cuttings in early summer

Briza media
Quaking/trembling grass/Doddering dillies

⬆ 60cm/24in ⬌ 30cm/12in   **EASY**

This fabulous small, loose, tufted perennial from Europe and Asia makes neat little hummocks of fresh green. It has fine-bladed grassy leaves and produces scaly pendent panicles that are first green, then wine coloured, ageing to buff as they mature. It thrives on next to nothing, does equally well in dry and damp spots and is unfussy about soil pH. A soft breeze will send the panicles all a-quiver, hence the common names. As grasses go, it is charming, delicate and tactile, and makes a delightful mist to show off other plants.

BEST USES Ideal in containers, but as happy studding a hot, dry, sunny border, planted with traditional herbaceous plants (looks terrific with *Hemerocallis*); handy in large swathes as ground cover; super for growing on difficult dry banks or rock and gravel gardens, and naturalises well

DROUGHT TOLERANCE Good, once established
FLOWERS June to August
SCENTED No
ASPECT Any, in a sheltered or exposed position; full sun
SOIL Any reasonable, well-drained soil
HARDINESS Fully hardy at temperatures down to -15°C/5°F; needs no winter protection
PROBLEMS None
CARE Trim lightly after flowering to maintain its compact habit
PROPAGATION Sow ripe seed in situ in mid-spring or early autumn; division in mid-spring to mid-summer

Calamintha grandiflora
Large calamint

⬆ 45cm/18in ⬌ 45cm/18in   **EASY**

This tough, low-growing, bushy herbaceous Mediterranean perennial (from the Pyrenees to Corsica) has downy, pale green foliage that smells deliciously of mint when crushed. It has tubular, streaky pink flowers and forms a pleasing green clump once established. Although it may not be a showstopper, it flowers reliably and is a pretty enough border plant.

BEST USES Very useful for ground cover in a lightly shaded woodland garden as it spreads readily; also a cottage garden favourite with both medicinal and culinary uses; butterflies and bees adore it, so ideal for the wildlife garden

DROUGHT TOLERANCE Good, once established
FLOWERS June to July
SCENTED Aromatic leaves
ASPECT South, west or east facing, in a sheltered or exposed position; full sun to partial shade
SOIL Any fertile, humus-rich, well-drained soil; add organic matter before planting; will struggle in waterlogged clay
HARDINESS Fully hardy at temperatures down to -15°C/5°F; needs no winter protection
PROBLEMS Powdery mildew
CARE Cut down to ground level in autumn or early spring
PROPAGATION Sow seed in a cold frame in spring; division in spring

Catananche caerulea
Cupid's dart

⬆ 60cm/24in ⬌ 30cm/12in   **EASY**

This pretty, deciduous, upright perennial from Italy and south-west Europe has multi-layered, papery, sky blue petals forming flowers about 5cm/2in across with contrasting dark centres from summer to early autumn. It looks very much like a cornflower and makes an excellent cut flower. The leaves are narrow, silver and almost grass-like, and it has very pretty silvery seed heads that rattle in light breezes.

> **BEST USES** Great in hot, sunny borders and especially good in gravel gardens; also good for wildflower gardens as it naturalises well with grasses and wildflowers, and attracts pollinating insects

DROUGHT TOLERANCE Good, once established
FLOWERS June to September
SCENTED No
ASPECT South, west and east facing, in a sheltered position; full sun
SOIL Any fertile, well-drained soil; will not thrive in waterlogged soils
HARDINESS Fully hardy at temperatures down to -15°C/5°F; needs no winter protection
PROBLEMS Powdery mildew
CARE Cut the plant back after flowering
PROPAGATION Sow seed at 15°C/59°F in spring; division in mid-spring; root cuttings in winter; or sow as an annual

Coreopsis 'Limerock Ruby'
Tickseed

⬆ 45cm/18in ⬌ 90cm/3ft   **EASY**

The flowers of coreopsis are usually yellow but this bushy, clump-forming herbaceous perennial from the USA is a rare red-flowered example. It has an upright habit with mid-green lance-shaped leaves and a profusion of ragged, ruby red flowers with orangey eyes, delicately edged with gold, in summer.

> **BEST USES** Good value in naturalised borders and cottage gardens as they are long lasting and low maintenance; ideal for wildflower gardens as they attract bees and butterflies

DROUGHT TOLERANCE Good, once established
FLOWERS June to September
SCENTED No
ASPECT South, west or east facing, in a sheltered or exposed position; full sun
SOIL Any well-drained, fertile soil
HARDINESS Fully hardy at temperatures down to -15°C/5°F; needs no winter protection
PROBLEMS Slugs and snails
CARE Deadhead to prolong flowering; cut old or damaged stems back to the base in autumn
PROPAGATION Sow seed at 10°C/50°F in spring; division in spring; basal stem cuttings in spring

Echinops ritro 🏅
Globe thistle

↕ 60cm/24in ↔ 45cm/18in   **EASY**

An increasingly popular, compact, upright herbaceous perennial from dry, rocky regions of Europe, the globe thistle has stiff, prickly, grey-green leaves and the most attractive globular flower heads that have a metallic sheen before maturing to a startling bright blue. The flowers are long lasting and very effective in flower arrangements when dried. Considering it thrives in the hungriest of soils, the globe thistle has a remarkable sculptural quality. With its erect structural shape, it has real understated architectural glamour.

BEST USES Stunning in a gravel or Mediterranean garden and would enhance a cottage garden or herbaceous border; naturalises well in traditional borders and wildflower schemes and encourages pollinating insects

DROUGHT TOLERANCE Excellent, once established

FLOWERS July to August

SCENTED No

ASPECT South, west or east facing, in a sheltered or exposed position; full sun

SOIL Any poor to moderately fertile, well-drained soil

HARDINESS Fully hardy at temperatures down to -15°C/5°F; needs no winter protection

PROBLEMS Aphids

CARE Deadhead to avoid over-zealous self-seeding; cut down to ground level in early spring

PROPAGATION Sow seed at 15°C/59°F in spring; root cuttings in late autumn

Festuca gautieri
(formerly *F. scoparia*) Bearskin fescue

↕ 15cm/6in ↔ 25cm/10in   **EASY**

A slightly unusual, rhizomatous, evergreen perennial grass from southern Europe, with a compact habit that forms pleasing, tactile, low mounds or rounded hummocks. The stiff, needle-like linear leaves are a rich green and in summer yellow-orange-tinted panicles float above the foliage. When clipped of its flowers, it hugs the ground, resembling a small green pouffe or prickly hedgehog.

BEST USES Very useful for edging beds, borders and gravel gardens; looks superb planted in multiples to create low-growing, low-maintenance ground cover in a contemporary garden; very striking planted in containers

DROUGHT TOLERANCE Excellent, once established

FLOWERS May to June, but grown mainly for foliage

SCENTED No

ASPECT Any, in a sheltered or exposed position; full sun to partial shade

SOIL Any fertile, well-drained soil, preferably slightly acid

HARDINESS Fully hardy at temperatures down to -15°C/5°F; needs no winter protection

PROBLEMS Rust

CARE Trim lightly once a year; divide overcrowded clumps every three years

PROPAGATION Sow ripe seed at a minimum of 10°C/50°F in spring or autumn; division in autumn

Festuca glauca 'Elijah Blue'
Blue fescue

⬆ 10cm/4in ↔ 10cm/4in   **EASY**

One of the loveliest evergreen grasses from south-west Europe, *Festuca glauca* 'Elijah Blue' has silver, almost metallic blue, narrow, thin blades that make small, low, very textural hummocks. The small, bluish panicles in summer turn to beige with age. A hot spot really brings out the blue hue of the leaves, and it looks great planted with moody black or purply plum-coloured flowers, or bright yellows and oranges. Definitely one of the best grass foliage plants around.

BEST USES I've seen this versatile grass used to great effect in containers, planted as a dwarf edging in a Mediterranean garden, or just dotted through a gravelled area and it always looks terrific; excellent for contemporary gardens

DROUGHT TOLERANCE Excellent, once established
FLOWERS June to July
SCENTED No
ASPECT Any, in a sheltered or exposed position; full sun
SOIL Any poor to moderately fertile, well-drained soil
HARDINESS Fully hardy at temperatures down to -15°C/5°F; needs no winter protection
PROBLEMS None
CARE Rake your fingers through the grass in winter and the dead leaves come away quite easily; may need to be divided every three years or so, as can become tatty with age
PROPAGATION Sow seed at 10°C/50°F in spring; division in autumn

Geranium endressii 🎖

⬆ 50cm/20in ↔ 90cm/3ft   **EASY**

This semi-evergreen, herbaceous hardy geranium from the French Pyrenees forms pleasing low clumps of neat, lobed, mid-green leaves and simple, shell pink flowers, with dainty lilac veining, from summer to autumn. It has a long and reliable flowering period, the first flush lasting some weeks, with a second display in September.

BEST USES Excellent in the herbaceous border, and as ground cover on awkward slopes, in light woodland, cottage gardens and wildflower areas

DROUGHT TOLERANCE Good, once established
FLOWERS May to September
SCENTED No
ASPECT Any, in a sheltered or exposed position; full sun to partial shade
SOIL Any fertile, well-drained soil; add organic matter before planting; will struggle in waterlogged clay
HARDINESS Fully hardy at temperatures down to -15°C/5°F; needs no winter protection
PROBLEMS Capsid bug and vine weevil; powdery mildew
CARE Cut back to ground level in early spring to promote new growth; cut back faded flower stems immediately after flowering for a second flush of flowers in autumn; divide overcrowded clumps in early spring or late summer or late autumn
PROPAGATION Sow ripe seed at 15°C/59°F in spring; basal stem cuttings in spring; division in early spring or early autumn (after flowering)

Helianthemum 'Wisley Primrose' ♙
Rock rose

⬆ 30cm/12in ↔ 45cm/18in   **EASY**

Helianthemum comes from the Latin and means 'sun flower,' which should give all the clues you need when siting this alpine. It comes from widespread locations across the world but mostly we grow garden hybrids. This pretty, woody, evergreen, spreading shrub has silvery grey leaves and puts out gloriously abundant blooms of papery, pale yellow flowers that darken to sherbet yellow towards the centre. Each flower lasts only a day, but flowering continues non-stop over a three-month period, during which the plant is smothered with fresh flower buds daily.

BEST USES A godsend planted en masse on a hot, dry bank in the full blast of the sun; also ideal in a traditional border or cottage garden

DROUGHT TOLERANCE Excellent, once established
FLOWERS May to September
SCENTED No
ASPECT South, west or east facing, in a sheltered or exposed position; full sun
SOIL Any fertile, well-drained soil; prefers neutral to alkaline soil
HARDINESS Fully hardy at temperatures down to -15°C/5°F; needs no winter protection
PROBLEMS None
CARE Trim flowering shoots lightly after flowering to prevent legginess
PROPAGATION Sow seed in spring in frost-free area; greenwood cuttings in summer or early autumn

Helichrysum italicum ♙
Curry plant

⬆ 60cm/24in ↔ 80cm/32in   **EASY**

This very aromatic dense, evergreen sub-shrub from the southern Mediterranean has small, narrow, silver-grey leaves, which smell of pungent curry spices when bruised. The foliage whitens as the days grow hotter and drier. It bears bobbly, light mustard-coloured flower heads that last from summer to early autumn and is an invaluable foliage plant for a dry, warm spot.

BEST USES Perfect as a backdrop plant for Mediterranean borders and gravel gardens, with its soft silvery leaves planted against deep mauves, reds and oranges; adds year-round interest to the herb garden

DROUGHT TOLERANCE Excellent, once established
FLOWERS July to September
SCENTED Aromatic leaves
ASPECT South, west or east facing, in a sheltered position with protection from cold winds; full sun
SOIL Any fertile, well-drained soil
HARDINESS Frost hardy at temperatures down to -5°C/23°F; needs winter protection in colder areas
PROBLEMS None
CARE Cut back flowered shoots to 2.5cm/1in of old growth if plant gets leggy, after all danger of frosts has passed
PROPAGATION Sow seed at 13–16°C/55–61°F in spring; heeled semi-ripe cuttings in summer

GREENFINGER TIP *Plant on a small ridge, to ensure good drainage*

Helichrysum petiolare ☻
Liquorice plant

⬆ 60cm/24in ⬌ 80cm/32in   **EASY**

This trailing evergreen foliage shrub from South Africa is used widely in summer hanging baskets. It has a branching habit with very attractive oval to heart-shaped leaves of the softest woolly grey, and bears small, off-white, insignificant flowers. It is not reliably hardy, so may need over-wintering in a greenhouse. *H.p.* 'Limelight' ☻ is a more unusual and very fetching variety with pale lime leaves.

BEST USES Perfect as a backdrop plant, with its soft, silvery leaves; an effective foliage plant for gravel gardens, containers and hanging baskets

DROUGHT TOLERANCE Excellent, once established
FLOWERS Insignificant; grown mainly for foliage
SCENTED Spice-scented leaves
ASPECT South or west facing, in a sheltered position with protection from cold winds; full sun
SOIL Any fertile, well-drained soil
HARDINESS Half-hardy at temperatures down to 0°C/32°F; may need winter protection
PROBLEMS None
CARE Remove straggling stems and damaged growth in spring; if plant gets leggy, cut back hard around May, after all danger of frosts has passed
PROPAGATION Sow seed at 13–16°C/55–61°F in spring; semi-ripe cuttings in summer

••

GREENFINGER TIP *Friends grow this successfully without winter protection in city gardens, year in, year out, so it is well worth chancing in mild areas*

Heuchera 'Obsidian'
Coral flower

⬆ 45cm/18in ⬌ 30cm/12in   **EASY**

Heucheras are clump-forming perennials, mainly from North America, and this evergreen variety has to be the cream of the crop. It has gorgeous, scallop-edged, deep purple to black (or as near as nature will allow) polished leaves that darken with age. Slender, upright, deep plummy stems hold multiple sprays of tiny, delicate, creamy white flowers from early summer: 10/10.

BEST USES Great as foliage plants in a formal or cottage garden; work well as ground cover or underplanting in a mixed or shrub border; also brilliant in pots or containers in a shady garden

DROUGHT TOLERANCE Excellent, once established
FLOWERS June to August
SCENTED No
ASPECT South, west or east facing, in a sheltered position; full sun to partial shade
SOIL Any fertile, humus-rich, well-drained soil; add organic matter before planting
HARDINESS Fully hardy at temperatures down to -15°C/5°F; needs no winter protection
PROBLEMS None
CARE Cut back foliage in early spring; mulch with organic matter around the crown in spring
PROPAGATION Division in autumn or every three years as colour and vigour decline with age

••

GREENFINGER TIP *A heuchera can heave itself out of the soil after frosts, so firm it down well in spring and mulch with organic matter*

Hordeum jubatum
Squirrel's tail grass

⬆ 50cm/20in ⬌ 30cm/12in **EASY**

This tufted short-lived perennial from Asia and North America is a delightful low-maintenance grass, with thin, narrow, linear, pale green leaves producing fine, delicate, silky light green bristles that mature to purple before fading to buff. It is a tactile plant and very pretty when its delicate flowers are stirred by the breeze. It self-seeds freely, so the longevity of the plants isn't really an issue, and is easily pulled out if it seeds itself where it is not wanted.

BEST USES Ideal as ground cover, as it will quickly cover bare ground; will grow well in containers; perfect in wildflower or prairie planting schemes

DROUGHT TOLERANCE Excellent, once established

FLOWERS June to July

SCENTED No

ASPECT South or west facing, in a sheltered position; full sun

SOIL Any fertile, well-drained soil

HARDINESS Fully hardy at temperatures down to -15°C/5°F; needs no winter protection

PROBLEMS None

CARE Low maintenance; cut back hard in spring

PROPAGATION Sow seed in situ in spring or autumn

Lavandula angustifolia 'Hidcote' ♀
English lavender

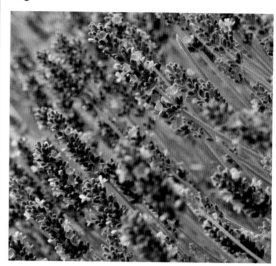

⬆ 60cm/24in ⬌ 75cm/30in **EASY**

This evergreen western Mediterranean shrub is one of the most enduringly popular of all the lavenders. It boasts a tidy, compact habit, and has heavenly scented deep blue flowers borne on narrow silver stems, rising from aromatic silver-grey foliage. Bees and butterflies adore this plant, adding a magical element to the garden as they hover above its fragrant purple haze. *L.a.* 'Miss Katherine' ♀ is a slightly larger variety with unusual, highly scented soft pink flowers and is one of the best pink cultivars.

BEST USES Perfect as a fragrant edging for paths and borders or as a small, informal hedge; ideal for Mediterranean and cottage gardens; handy for underplanting shrub roses as it clothes their naked lower limbs; also good in coastal gardens

DROUGHT TOLERANCE Excellent, once established

FLOWERS July to September

SCENTED Scented flowers and aromatic foliage

ASPECT South, west or east facing, in a sheltered position; full sun

SOIL Any fertile, well-drained soil

HARDINESS Fully hardy at temperatures down to -15°C/5°F; needs no winter protection

PROBLEMS Froghoppers; botrytis (grey mould)

CARE Trim lightly in spring, taking care not to cut into the old wood; cut the flower stalks back to leaf level after flowering to prevent plant becoming leggy

PROPAGATION Sow seed in a cold frame in spring; semi-ripe cuttings in early summer

Lavandula stoechas
French lavender

↑ 60cm/24in ↔ 60cm/24in   **EASY**

This compact, evergreen, bushy shrub from the Mediterranean is becoming more popular as our milder climate has made it less hassle to grow. It has aromatic, linear silver leaves and wiry stems topped with striking, scented, deep purple flowers, with distinctive, incredibly dainty pale mauve bracts, resembling butterfly wings. Like all the lavenders, it is incredibly drought resistant.

BEST USES Perfect for the cottage garden, where bees and butterflies swarm around it; does well in containers and traditional herbaceous borders

DROUGHT TOLERANCE Excellent, once established

FLOWERS July to September

SCENTED Scented flowers and aromatic foliage

ASPECT South or west facing, in a sheltered position; full sun

SOIL Any fertile, well-drained soil

HARDINESS Fully hardy/borderline at temperatures down to -15°C/5°F; may need winter protection in colder areas

PROBLEMS Froghoppers; botrytis (grey mould)

CARE Low maintenance; cut away the faded flower stalks to leaf level to keep growth bushy and compact

PROPAGATION Sow seed in a cold frame in spring; semi-ripe cuttings in early summer

••

GREENFINGER TIP *Although French lavender is less hardy than other lavenders, city dwellers and people living in mild areas find this often comes through average winters without extra protection*

Lavandula viridis
Green lavender

↑ 60cm/24in ↔ 60cm/24in   **EASY**

This unusual lavender from Portugal and southern Spain is startlingly different. It is a very tidy, compact, evergreen shrub, with narrow, light green, aromatic foliage, smooth hairless stems and fat flower heads of greenish yellow, topped with delicate pale lemon flags that have a citrus scent. The flowers excite curious enquiry and admiration in equal measure.

BEST USES Ideal for a sheltered, sunny border in a Mediterranean, gravel or coastal garden; does well in warm city patios and courtyards; excellent as a container plant if drainage is good

DROUGHT TOLERANCE Excellent, once established

FLOWERS June to July

SCENTED Scented flowers and aromatic foliage

ASPECT South or west facing, in a sheltered position; full sun

SOIL Any fertile, well-drained soil

HARDINESS Frost hardy down to –5C/23°F; may need winter protection in colder areas

PROBLEMS Froghoppers; botrytis (grey mould)

CARE Low maintenance; cut away the faded flower stalks to leaf level to keep growth bushy and compact

PROPAGATION Sow seed in a cold frame in spring; semi-ripe cuttings in early summer

••

GREENFINGER TIP *All lavenders thrive on very well-drained, poor soil; if your soil is not free draining, grow plants on a small ridge for sharp drainage and they will flourish*

Origanum vulgare
Wild marjoram

⬆ 30cm/12in ⬌ 30cm/12in   **EASY**

The name '*vulgare*' may suggest this rhizomatous, woody-based, bushy perennial is unremarkable. On the contrary! It has small, rounded, very aromatic green leaves that can be used as flavouring for food and salads but also forms spreading stems, jewelled with masses of pretty, tubular, light to dark pink flowers throughout the summer to early autumn. It is seen growing wild along roads and lanes in France and Europe, so that gives an indication of how low maintenance and drought resistant it is.

BEST USES A great asset in the herb garden, and will do well in containers by the kitchen door; also suitable for cottage gardens and rockeries, and is very ornamental in beds and borders

DROUGHT TOLERANCE Excellent, once established
FLOWERS July to September
SCENTED Aromatic leaves
ASPECT Any, in a sheltered or exposed position; full sun
SOIL Any fertile, well-drained soil
HARDINESS Fully hardy at temperatures down to -15°C/5°F; needs no winter protection
PROBLEMS None
CARE Cut back in late summer to keep plant tidy
PROPAGATION Sow seed in a cold frame in autumn (germination can be variable); division in spring after flowering; softwood cuttings in summer

Potentilla × *tonguei* 🎖
Cinquefoil

⬆ 10cm/4in ⬌ 30cm/12in   **EASY**

Many of the cinquefoils hail from rocky mountains to meadows and do well under dry conditions if bolstered with organic matter, and this is one of the more naturally drought-tolerant varieties. This garden hybrid is a low-growing, clump-forming herbaceous perennial with dark green palmate leaves and produces vivid, shallow, saucer-shaped apricot flowers with deep orangey red centres in fair abundance through summer. It is a cheerful, tough, versatile plant.

BEST USES Grow as carpeting cover in a gravel garden; blends easily in cottage gardens or studding the front of the summer border; does well in containers so ideal for bringing exotic colour into a small sunny courtyard or patio

DROUGHT TOLERANCE Good, once established
FLOWERS June to August
SCENTED No
ASPECT South or west facing, in a sheltered position; full sun
SOIL Any fertile, humus-rich, well-drained garden soil; add organic matter before planting
HARDINESS Fully hardy at temperatures down to -15°C/5°F; needs no winter protection
PROBLEMS None
CARE Mulch with organic matter in early spring; cut back flowering stems after flowering to 2.5cm/1in of old growth
PROPAGATION Sow seed in a cold frame in spring; division in spring or autumn

Scabiosa 'Butterfly Blue'
Pincushion flower

↑ 40cm/16in ↔ 40cm/16in   **EASY**

Scabious are herbaceous perennials from the rocky areas of the Mediterranean, and their pincushion-like flowers are a familiar and much-loved mainstay of the garden, with their long flowering period and ability to thrive with minimal care. This variety has narrow, pale grey, lance-shaped leaves that clump up nicely and slender stems topped with very pretty, dainty lavender-blue flowers with frilled edges and paler lilac pincushion centres. I have seen it in flower right up to the first frosts in a mild autumn.

BEST USES Enduringly popular in the cottage garden, and much coveted by bees, butterflies and hoverflies, so a natural choice for the flowering meadow or wildflower garden; also a good cut flower

DROUGHT TOLERANCE Excellent, once established
FLOWERS July to September
SCENTED No
ASPECT South or west facing, in a sheltered or exposed position; full sun to partial shade
SOIL Any fertile, well-drained soil; will not do well in winter wet
HARDINESS Fully hardy at temperatures down to -15°C/5°F; needs no winter protection
PROBLEMS Powdery mildew
CARE Deadhead flower stems to encourage further flowering; cut back to the ground in autumn
PROPAGATION Sow ripe seed in situ or in a cold frame in spring; division in spring

Sempervivum 'Commander Hay' ♀
House leek

↑ 8cm/3in ↔ 30cm/12in   **EASY**

Sempervivums are small, evergreen, succulent perennials, from the mountains of Europe and Asia. An odd little package really, with their green rosettes and sharp leaf margins that form thick, cobwebbed mats. Leaf colour varies from bright green to purple and grey, and the flower colours vary from red to white and yellow. This variety has shiny, wine-coloured leaves with grey-green tips and small, greenish flowers in summer.

BEST USES Grow on a sunny patio in an old enamel sink or trough, where they get the benefit of the sharp drainage; use to plug gaps in walls

DROUGHT TOLERANCE Excellent, once established
FLOWERS Insignificant; grown mainly for foliage
SCENTED No
ASPECT South, west or east facing, in a sheltered or exposed position; full sun
SOIL Any very well-drained, gritty soil
HARDINESS Fully hardy at temperatures down to -15°C/5°F; needs no winter protection
PROBLEMS Rust
CARE Low maintenance; rosettes die back after flowering
PROPAGATION Sow seed in pots in a cold frame in spring; root offsets in summer to autumn

GREENFINGER TIP *Using a clean sharp knife, cut away one of the offsets arranged like small pincushions at the base of the main plant, pot up in gritty compost, and you have a new plant*

Stachys byzantina
Lambs' ears

⬆ 45cm/18in ⬌ 60cm/24in　　　**EASY**

I first came across this evergreen hardy perennial in George Bernard Shaw's garden (though it's originally from the Caucasus and Iran) and was enchanted by the tactile, downy rosettes of velvety soft, grey leaves that are thickly covered in soft white wool and form a dense mat. Stiff, upright spikes of purple flowers are borne in summer to early autumn in long, hot summers, and are an irresistible lure to bees and butterflies. *S.b.* 'Silver Carpet' is a non-flowering variety that makes unbeatable tough ground cover.

BEST USES Ideal for ground cover in a hot sunny border, Mediterranean or gravel garden; also pretty in a cottage garden and attractive to pollinating insects; perfect for kid gardeners, who love the soft 'lambs' ears'

DROUGHT TOLERANCE Excellent, once established

FLOWERS June to September

SCENTED No

ASPECT South, west or east facing, in a sheltered or exposed position; full sun

SOIL Any fertile, well-drained soil

HARDINESS Fully hardy at temperatures down to -15°C/5°F; needs no winter protection

PROBLEMS Powdery mildew

CARE Remove faded flower stems

PROPAGATION Sow seed in a cold frame in spring or autumn; division in spring, potting up rooted clumps immediately

Thymus pulegioides 'Archer's Gold'
(formerly *T. citriodorus* 'Archer's Gold')

⬆ 25cm/10in ⬌ 45cm/18in　　　**EASY**

This is a tidy, creeping sub-shrub, making an evergreen mound of fragrant matting. Its golden-variegated leaves smell deliciously lemony when crushed underfoot, giving the plant its common name of lemon-scented thyme, and are tough enough to withstand foot traffic without suffering any real harm. Myriad small pale lilac flowers, held on short stems, jewel the foliage in summer. Traditionally, thymes were used for their antiseptic properties and are popularly used to flavour cooking.

BEST USES A lovely plant for the herb pot or herb garden; works very well as edging at the front of a border and makes excellent ground cover; charming planted in cracks between paving stones or left to cover low stone walls

DROUGHT TOLERANCE Good, once established

FLOWERS June to September

SCENTED Aromatic leaves

ASPECT South, west or east facing, in a sheltered position; full sun

SOIL Any fertile, well-drained soil, especially neutral to alkaline soils

HARDINESS Fully hardy at temperatures down to -15°C/5°F; needs no winter protection

PROBLEMS None

CARE A light trim after flowering will keep leaves compact

PROPAGATION Division in spring; softwood cuttings in early summer

Agapanthus 'Black Pantha'
African lily

⬆ 90cm/3ft ⬌ 45cm/18in   **EASY**

I'm a pushover for dark-coloured blooms and this sumptuous, evergreen perennial from South Africa, with fleshy, tuberous roots, has myriad tiny tubular flowers that form large globe flower heads of the most delicious purply black. Once the flowers are spent, the silvery seed heads add skeletal architectural form to the border in October and often last all the way through to November. For those bothered by furry intruders, agapanthus are reputedly rabbit proof.

BEST USES Very effective in drifts for height in the Mediterranean garden or summer border; ideal for containers; should do well in coastal areas

DROUGHT TOLERANCE Excellent, once established

FLOWERS June to July

SCENTED No

ASPECT South or west facing, in a sheltered position; full sun

SOIL Any fertile, well-drained soil

HARDINESS Fully hardy/borderline at temperatures down to -15°C/5°F; may need winter protection in colder areas

PROBLEMS Slugs and snails

CARE Mulch the crown in winter to keep it dry and protect it from frosts

PROPAGATION Division in spring

••

GREENFINGER TIP *You will need a tough tool to cut through the roots when dividing this plant: I find an old billhook does the trick nicely!*

Allium sphaerocephalon
Round-headed leek

⬆ 90cm/3ft ⬌ 10cm/4in   **EASY**

This summer-flowering bulbous perennial, originally from areas spanning Europe, north Africa and Asia, produces compact, cone-shaped, wine-coloured flower heads on tall, bare, erect stems with linear strappy leaves. The leaves may scorch in prolonged periods of drought, but don't let that worry you: they will be hidden by neighbouring plants. Although not as showy as other varieties, it looks very classy with muted grey foliage or grasses.

BEST USES An understated plant for the gravel garden or hot, dry border; useful in the herbaceous border and cottage garden, and naturalised with ornamental grasses; attracts pollinating insects, so an excellent subject for the wildflower garden

DROUGHT TOLERANCE Good, once established

FLOWERS July to August

SCENTED Onion-scented leaves when bruised

ASPECT South, west or east facing, in a sheltered position; full sun

SOIL Any fertile, well-drained soil

HARDINESS Fully hardy at temperatures down to -15°C/5°F; needs no winter protection

PROBLEMS Onion white rot and powdery mildew

CARE Mark the spot where they are planted as the leaves die back in winter; cover the crowns with dry leaves or old fern fronds in winter to encourage reliable flowering

PROPAGATION Sow ripe seed in a cold frame in spring; bulbils in late summer

Alstroemeria aurea
Peruvian lily

⬆ 90cm/3ft ↔ 90cm/3ft   **MEDIUM**

Peruvian lilies have a reputation for being difficult to establish and invasive once they settle down, but many of the new cultivars have a more reliable constitution, with longer flowering periods. Dainty from top to toe, but with tough, tuberous roots, this perennial has long, narrow, mid-green lance-shaped leaves and very slender, fairly erect flower stems, with showy, yellow-gold lily-like flowers with chocolate markings.

BEST USES Grow in drifts or groups along a formal flower border or in a cottage garden; will do well in containers in a city or patio garden

DROUGHT TOLERANCE Good, once established
FLOWERS July to September
SCENTED No
ASPECT South, west or east facing, in a sheltered position; full sun to partial shade
SOIL Any fertile, moist, well-drained soil; add organic matter before planting
HARDINESS Frost hardy at temperatures down to -10°C/14°F; can withstand brief drops in temperature down to -15°C/5°F but needs protection from winter frosts in cold regions
PROBLEMS Slugs eat young foliage; red spider mite if grown indoors or in a greenhouse
CARE Mulch the crown in winter to protect it from frosts; divide overcrowded clumps in autumn (they resent root disturbance, so need careful handling followed by mulching)
PROPAGATION Division in late summer or autumn

Artemisia abrotanum 🎖
Lad's love/Southernwood/Maiden's ruin

⬆ 75cm/30in ↔ 50cm/20in   **EASY**

A southern European shrub that is deciduous in colder regions or semi-evergreen in milder areas. It has a compact, fairly upright habit, with tactile, ferny, velvet grey-green leaves that are very aromatic when bruised. It bears small yellow or dirty white button-like flowers in summer, but these can be trimmed off: it is grown mainly for its appealing foliage. It is reputed to be a marvellous insect repellent.

BEST USES Best in a Mediterranean border or cottage garden: complements poppies, lilies, grasses and *Helianthemum* (rock rose) incredibly well

DROUGHT TOLERANCE Excellent, once established
FLOWERS August, but grown mainly for foliage
SCENTED Aromatic leaves
ASPECT South, west or east facing, in a sheltered or exposed position; full sun
SOIL Any fertile, well-drained soil; hates sitting in winter wet and can be short-lived in heavy wet soil
HARDINESS Fully hardy at temperatures down to -15°C/5°F; needs no winter protection
PROBLEMS Aphids
CARE Cut back in spring to prevent a tendency to legginess; pinch out new shoots to keep the plant bushy and encourage a finer foliage display
PROPAGATION Division in spring; take regular heeled greenwood cuttings in mid to late summer

Brachyglottis (Dunedin Group) 'Sunshine' Senecio 'Sunshine'

↑ 90cm/3ft ↔ 1.5m/5ft **EASY**

A tough evergreen shrub from New Zealand that has a naturally mounding habit. The oval, silver-grey leaves have white felty undersides and masses of shaggy, bright sunshine yellow, daisy-like flowers are produced in summer. These are a rather brash yellow, and it can be worth trimming the shrub of its flowers to bring out the more attractive foliage, which appears whiter in the midday sun.

BEST USES Ideal in hot, sunny borders; very tolerant of coastal conditions, so will do well for gardens exposed to salt-laden winds

DROUGHT TOLERANCE Excellent, once established
FLOWERS June to July
SCENTED No
ASPECT South or west facing, in a sheltered or exposed position; full sun to partial shade (but flowering is reduced and the leaf colour less pronounced in shade)
SOIL Any fertile, moist, well-drained soil; add organic matter before planting
HARDINESS Fully hardy at temperatures down to -15°C/5°F; needs no winter protection
PROBLEMS None
CARE Trim back lightly after flowering to maintain a bushy appearance
PROPAGATION Semi-ripe cuttings in summer

Caryopteris × *clandonensis* 'Heavenly Blue'

↑ 90cm/3ft ↔ 90cm/3ft **EASY**

This bushy, upright, deciduous shrub from mountainous Asia is a firm favourite, valued for its tactile, aromatic (a little like lavender), narrow, pointed grey-green foliage, which is reason enough to grow it. The leaves provide the perfect foil for the brilliant profusion of lightly fragrant, bright azure blue flowers in summer to early autumn. It is a lovely small shrub that really earns its keep in the late summer border, and this variety has one of the bluest flowers of them all.

BEST USES A lovely addition to the summer flower border, and bees and butterflies find it very appealing, so a great choice for wildlife gardens; looks very becoming planted as a backdrop to low-growing architectural grasses or underplanted with summer bulbs such as galtonia and acidanthera

DROUGHT TOLERANCE Good, once established
FLOWERS July to September
SCENTED Aromatic leaves and flowers
ASPECT South or west facing, in a sheltered position; full sun
SOIL Any fertile, well-drained soil
HARDINESS Fully hardy at temperatures down to -15°C/5°F; needs no winter protection
PROBLEMS Capsid bug
CARE In spring, cut the flowering stems back to the old framework
PROPAGATION Softwood cuttings in spring; greenwood cuttings in early summer

Dianthus barbatus
Sweet William

⬆ 75cm/30in ⬌ 30cm/12in   **EASY**

People often think of members of the carnation and 'pinks' family as being dainty little things, but in reality they don't come much tougher. Sweet Williams are very popular, old-fashioned evergreen herbaceous perennials, but they are short-lived, diminishing from their second year, so are best treated as biennials. This popular variety with light green leaves bears small, single, sweetly clove-scented flowers in clusters, in varying hues of purple and pink with whitish centres. Their long-lived foliage and long flowering period have made them a great favourite in cottage gardens.

BEST USES Great at front of borders, and an excellent cut flower; thrive in containers; attractive to pollinating insects, so good for the wildflower garden

DROUGHT TOLERANCE Excellent, once established
FLOWERS June to September
SCENTED Scented flowers
ASPECT South, west or east facing, in a sheltered or exposed position; full sun
SOIL Any fertile, well-drained soil
HARDINESS Fully hardy at temperatures down to -15°C/5°F; needs no winter protection
PROBLEMS Aphids; rust; can be fairly short-lived
CARE Trim non-flowering shoots in summer; deadhead spent flowers
PROPAGATION Sow seed in a cold frame in early summer; semi-ripe cuttings in mid to late summer

Eryngium × tripartitum 🎖
Sea holly

⬆ 90cm/3ft ⬌ 50cm/20in   **EASY**

An architectural taprooted perennial that makes sculptural clumps of spiny grey-green leaves and branching stems, capped in summer with stiff flower heads of the most vivid blue steel, shaped rather like teasel. One look at it tells you it's a Mediterranean plant, with glaucous, sinewy toughness and the colour at its best in full sun. A low-maintenance plant, it looks well through to late winter covered in a dusting of frost (and the dried flowers can be used in floral arrangements).

BEST USES An absolute must for the gravel or Mediterranean garden; also useful in cottage and wildflower gardens as it attracts pollinating insects; will do well for coastal gardeners

DROUGHT TOLERANCE Excellent, once established
FLOWERS July to August
SCENTED No
ASPECT South, west or east facing, in a sheltered or exposed position; full sun
SOIL Any poor to moderately fertile, well-drained soil
HARDINESS Fully hardy at temperatures down to -15°C/5°F; needs no winter protection
PROBLEMS Powdery mildew and root rot
CARE Cut down to ground level in spring
PROPAGATION Sow ripe seed at 10°C/50°F in spring; division in spring; root cuttings in late autumn

GREENFINGER TIP *They will not tolerate wet: this chap is a sun-worshipper, and doesn't like to get his feet drenched!*

Hemerocallis 'Bela Lugosi'
Day lily

⬆ 80cm/32in ⬌ 60cm/24in   **EASY**

Hemerocallis are clump-forming perennials from Asia and this spectacular modern cultivar is absolutely corking. It has great foliage: the fresh, texturally interesting, strap-like leaves are a pleasing lettuce green, and look good most of the year. The dramatic, scented flowers are large (15cm/6in), sumptuous and figgy wine-coloured, with slightly crimped edges and striking contrasting gold and plum stamens, rising from a greeny yellow throat. (I am informed by the ever-knowledgeable Mr Loftus of Woottens Plants that Bela Lugosi, an actor famous for his Dracula roles, was buried in full film costume regalia!)

BEST USES Splendid planted amongst grasses, and naturalises well; also fabulous dotted in clumps through a traditional herbaceous border

DROUGHT TOLERANCE Excellent, once established
FLOWERS July
SCENTED Lightly scented flowers
ASPECT South, west or east facing, in a sheltered or exposed position; full sun
SOIL Any fertile, moist, well-drained soil; add organic matter before planting
HARDINESS Fully hardy at temperatures down to -15°C/5°F; needs no winter protection
PROBLEMS None, though flower colour may fade in direct overhead sunlight
CARE Deadhead fading flowers; foliage will die back
PROPAGATION Left undisturbed, hemerocallis will colonise beautifully; division in early spring or autumn

Knautia macedonica

⬆ 80cm/32in ⬌ 45cm/18in   **EASY**

This unassuming clump-forming herbaceous perennial from the Balkans is a simple plant with a natural, sprawling habit. The lobed green leaves and slender, slightly branching stems are topped with small, but very free-flowering, bristly, button-like flowers, the colour of mulled wine, that colour up best in full sun. Invaluable for adding interest to the late summer border, it is low maintenance and, though relatively common, delivers colourful cheer by the spadeful.

BEST USES Marvellous for the wildflower or cottage garden, as bees and insects love it; the pretty flower heads can be best appreciated when grown amongst grasses

DROUGHT TOLERANCE Excellent, once established
FLOWERS July to September
SCENTED No
ASPECT South, west or east facing, in a sheltered or exposed position; full sun to partial shade (but flowering is reduced in shade)
SOIL Any dry, fertile, well-drained soil
HARDINESS Fully hardy at temperatures down to -15°C/5°F; needs no winter protection
PROBLEMS Aphids
CARE Stake early in the year to prevent it splaying untidily; deadhead to prolong flowering
PROPAGATION Sow seed at 15°C/59°F in spring; division in spring; basal stem cuttings in spring

Lychnis coronaria ♀
Rose campion/Dusty miller

⬆ 80cm/32in ⬌ 45cm/18in   **EASY**

This short-lived herbaceous perennial from southern Europe is also grown as a biennial. It has soft, felty grey leaves and upright silver-grey stems bearing short branching stems of single, striking, vivid pink flowers in summer and well into early autumn. This is a tough low-maintenance plant that will very quickly dot pleasing magenta clumps of colour across the border.

BEST USES Ideal for prairie-style planting, cottage gardens and dry gravel or Mediterranean gardens; also indispensable for seaside gardens, withstanding coastal conditions well

DROUGHT TOLERANCE Excellent, once established

FLOWERS June to September

SCENTED No

ASPECT South, west or east facing, in a sheltered position; full sun to partial shade

SOIL Any dry, fertile, well-drained soil

HARDINESS Fully hardy at temperatures down to -15°C/5°F; needs no winter protection

PROBLEMS None

CARE Deadhead flowers to prevent over-zealous self-seeding

PROPAGATION Self-seeds easily; sow seed in a cold frame in spring; division in spring; basal stem cuttings in spring

Nepeta 'Six Hills Giant'
Catmint

⬆ 90cm/3ft ⬌ 60cm/24in   **EASY**

Both you and your cats will adore this informal, clump-forming herbaceous perennial from a wide variety of habitats across the northern hemisphere that has narrow, aromatic, pale grey-green leaves and tall spires of lavender-blue flowers in abundant profusion in mid-summer. I've seen this growing as the sole plant in a long avenue border with an evergreen hedge as a backdrop and it looked simply stunning.

BEST USES Marvellous for the nature garden, as bees and insects love it; looks handsome in swathes through a gravel garden or in a sunny courtyard border; cats adore nibbling the leaves

DROUGHT TOLERANCE Excellent, once established

FLOWERS June to July

SCENTED Aromatic leaves

ASPECT South, west or east facing, in a sheltered or exposed position; full sun to partial shade (but flowering is reduced in shade)

SOIL Any fertile, well-drained soil

HARDINESS Fully hardy at temperatures down to -15°C/5°F; needs no winter protection

PROBLEMS None, though may be affected by powdery mildew in long, hot, dry summers

CARE Provide supports early in the year, or support amongst sturdy grasses or perennials; trim after flowering to keep plant bushy

PROPAGATION Self-seeds freely; sow seed in a cold frame in autumn; division in spring or autumn

Oenothera fruticosa 'Fyrverkeri' 🏅
Evening primrose

⬆ 90cm/3ft ⬌ 30cm/12in   **EASY**

A tough, cheery and natural-looking North American herbaceous perennial. It has lance-shaped mid-green leaves with purplish tints and produces papery reddish buds that fall away, opening to lightly fragrant, cupped, buttercup yellow flowers during the summer months. The flowers don't open fully until dusk, when the perfume is strongest and attracts night-flying insects. It self-seeds happily and needs little maintenance.

BEST USES Will do equally well as prairie-style planting or in the cottage garden; will spread in colourful drifts through the gravel garden

DROUGHT TOLERANCE Excellent, once established

FLOWERS June to September

SCENTED Lightly scented flowers

ASPECT South or west facing, in a sheltered or exposed position; full sun

SOIL Any poor to fertile, stony, well-drained soil

HARDINESS Fully hardy at temperatures down to -15°C/5°F; needs no winter protection

PROBLEMS Powdery mildew

CARE Cut down after flowering

PROPAGATION Self-seeds easily; sow seed in a cold frame in spring; division in spring

Parahebe perfoliata 🏅
Diggers speedwell

⬆ 75cm/30in ⬌ 45cm /18in   **EASY**

This excellent Australian evergreen perennial is not nearly as popular as it should be. Perhaps it's the lax (some might say prostrate!) habit of the plant that is off-putting. It has attractive, rounded, green-grey leaves borne on long, low, arching, slender purple-blue stems. These are topped with slightly drooping, soft lavender-blue flower spikes in early to mid-summer, giving the charming impression that it is dozing in the midday sun.

BEST USES Marvellous for the nature garden, as bees and insects love it and their drowsy flight adds a magical dimension to any traditional or cottage garden; very effective as low ground cover in gravelled areas or on sloping banks

DROUGHT TOLERANCE Excellent, once established

FLOWERS June to July

SCENTED No

ASPECT South or west facing, in a sheltered position with protection from cold winds; full sun

SOIL Any fertile, well-drained soil

HARDINESS Frost hardy at temperatures down to -5°C/23°F; needs winter protection in colder areas, but the rootstock is fairly hardy, so top-growth damaged by severe cold will sprout again

PROBLEMS Slugs

CARE Cut back after flowering for a second, less profuse, flush of flowers in late summer; trim stems after flowering is over to keep plant bushy

PROPAGATION Sow seed in a cold frame in spring; semi-ripe cuttings in early spring or mid-summer

Phlomis fruticosa ♀

Jerusalem sage

⬆ 90cm/3ft ⬌ 1.5m/5ft   **EASY**

This is a simply lovely, low-maintenance aromatic evergreen shrub, originally from the Mediterranean. It is fairly upright in its growing habit and makes a very attractive mound of extremely tactile, soft, felty, grey-green sage-like leaves that look good over a long period. Intriguing hooded golden yellow flowers are produced from late spring to late summer. *P. russeliana* ♀ has pale yellow flowers and is more compact.

> **BEST USES** Looks at home in a hot-coloured sunny border or cottage garden; blends well with any blue or purple-flowering perennials and the linear foliage of *Crocosmia* (montbretia) suits it perfectly; should do well for coastal gardeners

DROUGHT TOLERANCE Excellent, once established

FLOWERS May to September

SCENTED Aromatic leaves

ASPECT South, west or east facing, in a sheltered position; full sun

SOIL Any fertile, well-drained soil

HARDINESS Fully hardy at temperatures down to -15°C/5°F; needs no winter protection

PROBLEMS Leafhoppers

CARE Trim back lightly after flowering

PROPAGATION Sow seed at minimum 13°C/54°F in spring; division in spring

....

GREENFINGER TIP *This is fairly vigorous and spreads once established, so give it plenty of space when planting*

Salvia argentea ♀

⬆ 90cm/3ft ⬌ 60cm/24in   **EASY**

This is a really gorgeous southern European herbaceous biennial or short-lived perennial with large, tactile, felty grey leaves that form a large rosette from the base of the plant in summer and look good from spring into autumn. They are covered in so much white wool that they look almost white in strong sunlight. It bears very tall (up to 90cm/3ft) but rather ordinary, branched, pinkish white flower spikes in summer. Since these weaken the plant and spoil the foliage, it may be best to remove the young flower spikes and use this as a magnificent foliage plant.

> **BEST USES** Marvellous studding the Mediterranean or gravel garden; excellent for coastal locations

DROUGHT TOLERANCE Excellent, once established

FLOWERS July to September, but grown mainly for foliage

SCENTED No

ASPECT South or west facing, in a sheltered position with protection from excessive winter wet; full sun to partial shade

SOIL Any fertile, sharply drained, dry soil

HARDINESS Frost hardy at temperatures down to -5°C/23°F; needs winter protection in colder areas

PROBLEMS Slugs and snails

CARE Protect from winter wet, as excessive rain can rot the plant

PROPAGATION Self-seeds easily; sow seed in a cold frame in summer

Sisyrinchium striatum

⬆ 90cm/3ft ⬌ 25cm/10in   **EASY**

This evergreen, clump-forming rhizomatous perennial from South America has long, fairly stiff green-grey leaves that form a fan (similar to iris, which is from the same family), and produces modest spires of buttery yellow, slightly cupped flowers arranged in clusters on the slender stems. *S.s.* 'Aunt May' is another lovely variety worth considering, with creamy margined leaves and pale yellow flowers. *S.* 'E.K. Balls' is a more compact variety with deep mauve flowers.

BEST USES Very effective as vertical accents in the gravel or Mediterranean garden; super combined with swaying grasses, as its more rigid stance complements their fluidity beautifully

DROUGHT TOLERANCE Excellent, once established
FLOWERS June to July
SCENTED No
ASPECT South or west facing, in a sheltered or exposed position; full sun
SOIL Any fertile, well-drained soil; hates winter wet
HARDINESS Fully hardy at temperatures down to -15°C/5°F; needs no winter protection
PROBLEMS None, though can be prone to root rot
CARE Deadhead to prevent self-seeding
PROPAGATION Sow seed in a cold frame in spring, summer or autumn; division in spring

Verbascum 'Helen Johnson'
Mullein

⬆ 90cm/3ft ⬌ 30cm/12in   **EASY**

Verbascums are found in dry, stony places in Europe, Africa and Asia. They are either short-lived perennials or biennials, with a tall, upright habit. This new introduction is a compact evergreen perennial whose colouring makes it stand out from the mullein crowd. It has rosettes of large, felty, grey-green leaves and tall soft woolly flower buds that open in summer to reveal overlapping, dark apricot saucer-shaped flowers studded up long flower stems. Self-sown seedlings are incredibly drought resistant. *V. chaixii* 'Album' ☿ is another stunning compact verbascum, with mauve-centred creamy flowers.

BEST USES Will colonise a gravel or prairie garden or hot Mediterranean border beautifully; good for a wildlife garden as bees love them

DROUGHT TOLERANCE Excellent, once established
FLOWERS July to August
SCENTED No
ASPECT South or west facing, in a sheltered or exposed position; full sun
SOIL Any poor to moderately fertile, well-drained soil; will struggle in very waterlogged soil
HARDINESS Fully hardy at temperatures down to -15°C/5°F; needs no winter protection
PROBLEMS Powdery mildew
CARE Deadhead to limit self-seeding if you don't want the colours diluted; cut back unsightly dead stems
PROPAGATION Self-seeds freely; sow seed in a cold frame in spring; division before growth in spring

Achillea filipendulina 'Cloth of Gold'
Yarrow

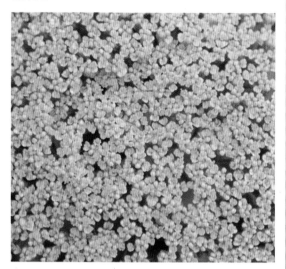

↑ 1.5m/5ft ↔ 60cm/24in   **EASY**

Yarrows are a group of upright, deciduous, hardy perennials from Eurasia with fine ferny foliage and flattened, plate-like flowers. This vigorous variety has large gold flower heads (13cm/5in across) from summer to autumn. They look stunning when frosted in winter, so resist the temptation to cut them down after flowering. *A.* 'Taygetea' has pale lemon flower heads fading to cream for those who find the bright yellows too brash. Yarrows need space to spread and do well.

BEST USES Good at the back of the border, planted with tall grasses, and in prairie plantings; very attractive to pollinating insects so a good choice for wildflower and cottage gardens

DROUGHT TOLERANCE Excellent, once established
FLOWERS June to September
SCENTED No
ASPECT South, west or east facing, in a sheltered or exposed position; full sun
SOIL Any fertile, well-drained soil; will struggle in waterlogged clay and dislikes sitting in winter wet
HARDINESS Fully hardy at temperatures down to -15°C/5°F; needs no winter protection
PROBLEMS Aphids; powdery mildew
CARE Cut back to ground level in spring; the flower heads are very heavy and absorb water like a sponge, so stake early in the year
PROPAGATION Sow seed in situ in spring; division in early spring

Aconitum napellus
Monkshood

↑ 1.5m/5ft ↔ 30cm/12in   **EASY**

A stately, upright deciduous herbaceous perennial from northern Europe, monkshood is a natural woodlander. This variety has deep green, toothed leaves and tall, elegant spires of rich purple flowers that are slightly hooded in appearance – hence the common name. *A.* 'Stainless Steel' with steely pale blue flowers is also a cracker.

BEST USES Lovely in a woodland or shady garden; good for wildflower gardens as it attracts pollinating insects; tolerant of coastal conditions

DROUGHT TOLERANCE Good, once established
FLOWERS July to August
SCENTED No
ASPECT West, north or east facing, in a sheltered position; full sun to partial shade
SOIL Any fertile, moist, well-drained soil; add organic matter before planting
HARDINESS Fully hardy at temperatures down to -15°C/5°F; needs no winter protection
PROBLEMS Powdery mildew
CARE Cut down to ground level in late winter or early spring; does not need staking
PROPAGATION Sow seed in a cold frame in spring (germination can be slow and erratic); division in early spring every three years (transplanted plants might sulk and take a year or so to revive)

GREENFINGER TIP *Every part of this plant is poisonous, so don't eat it!*

Alcea rosea 'Nigra'
Black hollyhock

⬆ 2m/6ft ↔ 60cm/24in   **EASY**

Never trust anyone who hates a hollyhock!
Looking at the huge variety on offer, it's easy to
see why they are so loved. I have seen a smart
urban courtyard planted with hollyhocks and
nothing else, and they looked divine. This short-
lived, tall-stemmed herbaceous perennial from
Asia has mid-green lobed leaves carried all the
way up the stems. The gorgeous large, chocolate-
brown blooms studding the main stem are a
definite lure for bees and nectar-seekers. They are
uncomplicated, always perform and do well on
meagre diets. Job done.

BEST USES A natural favourite for cottage and
wildlife gardens, as bees, butterflies and insects love
them; should do well in a coastal area

DROUGHT TOLERANCE Excellent, once established

FLOWERS June to July

SCENTED No

ASPECT South or west facing, in a sheltered position
with protection from cold winds; full sun

SOIL Any poor to fertile, well-drained soil

HARDINESS Fully hardy at temperatures down to
-15°C/5°F; needs no winter protection

PROBLEMS Rust can be a persistent nuisance

CARE Cut back to about 15cm/6in above ground level
in late autumn to early spring; may need staking as
strong winds will pull them over

PROPAGATION Self-seeds freely; sow seed at
15°C/60°F in late winter for annuals, or in situ in
summer for biennials

Baptisia australis 🎖
False indigo

⬆ 1.5m/5ft ↔ 60cm/24in   **EASY**

This is a large, spreading herbaceous perennial
from the USA that is very gracefully attired.
Fresh green three-ovalled leaflets form the leaves
and it carries a profusion of delicate, bright violet
blue lupin-like flowers on graceful, slender,
upright, green-grey glaucous stems that last well
as cut flowers. Puffy grey seed pods follow the
flowers. It's a wonder this isn't more popular: it
has all the stately elegance a summer border
could wish for.

BEST USES Plant to the rear of the flower border
or let it self-seed around a wildflower garden;
endlessly useful as ground cover clothing gently
sloping banks

DROUGHT TOLERANCE Excellent, once established

FLOWERS June to July

SCENTED No

ASPECT South, west or east facing, in a sheltered or
exposed position; full sun

SOIL Any fertile, well-drained soil, ideally sandy

HARDINESS Fully hardy at temperatures down to
-15°C/5°F; needs no winter protection

PROBLEMS None

CARE Deadhead to prevent self-seeding; cut back to
ground level once the foliage has browned

PROPAGATION Self-seeds freely; sow ripe seed in a
cold frame in spring; division in spring

Buddleja davidii 'Black Knight' ♀
Butterfly bush

⬆ 3m/10ft ⬌ 3m/10ft   EASY

Buddlejas originate from rocky areas of China and Japan. They thrive on a meagre diet, need little attention and grow like the clappers, though they can be relatively short-lived. This variety is as drought tolerant as they come. It is an arching deciduous shrub with mid-green lance-shaped leaves that yellow with age. It produces tiny, deep purple flowers that crowd together to form long, slightly drooping, nectar-rich fragrant flower spires, some 30cm/12in long, borne in splendid profusion in summer and attracting non-stop butterfly and bee traffic. *B. globosa* ♀ is worth a look, with pompoms of buttery orange flowers.

BEST USES Ideal for the natural or wildlife garden as bees and butterflies find them so attractive; I have used them as an informal flowering windbreak with great success

DROUGHT TOLERANCE Excellent, once established

FLOWERS July to September

SCENTED Sweetly scented flowers

ASPECT South, west or east facing, in a sheltered or exposed position; full sun

SOIL Any poor to fertile, well-drained soil

HARDINESS Fully hardy at temperatures down to -15°C/5°F; needs no winter protection

PROBLEMS None

CARE In early spring, prune stems hard within 3 buds of main framework; cut back flowering stems after flowering

PROPAGATION Semi-ripe cuttings in summer; hardwood cuttings in autumn

Callistemon citrinus 'Splendens' ♀
Crimson bottlebrush

⬆ 1.5m/5ft ⬌ 1.5m/5ft   EASY

Callistemons, originally from Australia, are increasingly popular as the climate becomes milder, and this variety is very striking indeed. It is a rounded evergreen shrub, with small, narrow, dark olive-coloured leaves and rather curious, exotic, soft, bristly, bottle brush-shaped flowers of deep crimson, followed by round seed pods. The flower heads are very tactile and attract constant traffic of hovering bees and butterflies.

BEST USES Ideal for the exotic garden; stunning as a swaying division between a Mediterranean garden and traditional parterre; thrive in containers

DROUGHT TOLERANCE Excellent, once established

FLOWERS June to July

SCENTED Aromatic foliage

ASPECT South or west facing, in a sheltered position with protection from cold winds; full sun

SOIL Any fertile, well-drained soil, especially neutral to acid soils

HARDINESS Frost tender at temperatures down to 5°C/41°F; needs winter/indoor protection in all but the mildest areas

PROBLEMS Mealybug and red spider mite, in a greenhouse

CARE Water well in the first year to establish; trim lightly after flowering and remove any damaged material; will tolerate hard pruning

PROPAGATION Surface-sow seed at a minimum constant temperature of 16°C/61°F in spring; semi-ripe cuttings in late summer

Cistus ladanifer 'Minstrel'
Common gum cistus

⬆ 2m/6ft ↔ 1.5m/5ft   **EASY**

Cistus come from dry, stony terrain in southern Europe and are renowned for their drought tolerance, but are no less showy or desirable for it. This common but desirable evergreen variety has aromatic, slightly sticky, dark green leaves that make a rounded shrub, and in summer it is studded with large, white, tissue-paper flowers, marked with a central maroon brown splash. The flowers only last a day, but numerous others magically appear overnight to replace them, so you never notice their brevity. It's a lovely thing.

> **BEST USES** Splendid on a dry rock bank, or in a hot gravel or Mediterranean garden; attractive as an informal flowering hedge in coastal areas

DROUGHT TOLERANCE Excellent, once established

FLOWERS June

SCENTED Aromatic leaves

ASPECT South or west facing, in a sheltered position with protection from cold winds; full sun

SOIL Any poor to moderately fertile, light, well-drained soil; chalky soil may make them chloritic (i.e. the leaves yellow)

HARDINESS Frost hardy at temperatures down to -5°C/23°F; needs winter protection in colder areas

PROBLEMS None

CARE Resents pruning, so pinch back the flowering shoots once they have faded

PROPAGATION Sow ripe seed in a cold frame in spring; softwood cuttings in summer

Crambe cordifolia ♀
Great sea kale

⬆ 2.5m/8ft ↔ 1.5m/5ft   **EASY**

I first saw this large, pretty, clump-forming herbaceous perennial from the Caucasus at Anglesey Abbey, planted alongside some fairly decadent dahlias, and it's a showstopper if you have the room. It has large bristly green leaves at the base and tall, thin, wiry green stems that are hidden by billowing froths of massed, scented, tiny white flowers that really catch your eye. The flowering time is short, but its sheer joyful abundance makes it worth every fleeting moment.

> **BEST USES** Ideal at the back of a hot sunny border and looks stunning with dark purples or hot colours; at home in informal wildflower and cottage gardens, and bees and insects love it

DROUGHT TOLERANCE Excellent, once established

FLOWERS June to July

SCENTED Lightly scented flowers

ASPECT South, west and east facing, in a sheltered position with protection from cold winds; full sun to partial shade

SOIL Most fertile, well-drained soils; can be short-lived on heavy soils

HARDINESS Fully hardy at temperatures down to -15°C/5°F; needs no winter protection

PROBLEMS The leaves grow tatty with age; slugs and snails may prove a nuisance in spring

CARE Cut back to ground level in autumn

PROPAGATION Sow seed at 10°C/50°F or in a cold frame in spring; division in spring; root cuttings in winter

Crocosmia 'Lucifer' ♟
Montbretia

⬆ 1.2m/4ft ↔ 1.2m/4ft   **EASY**

Originally from South Africa, montbretias are clump-forming perennials grown from corms. They have long, sword-like fresh green leaves and flowers predominantly in reds, oranges and yellows, opening progressively on tall, smooth, slender, slightly arching wiry stems. This is a cracking variety, with purple branching stems dangling with bright red flowers. The foliage is an appealing feature for nearly nine months of the year, being attractive when emerging in spring and architectural when mature.

BEST USES A favourite for a 'hot' border; excellent as vertical contrasts in any border; striking and unfussy in a contemporary garden; will do well in coastal areas; excellent for cutting

DROUGHT TOLERANCE Excellent, once established
FLOWERS August to September
SCENTED No
ASPECT South, west and east facing, in a sheltered or exposed position; full sun to partial shade
SOIL Any humus-rich, moist, well-drained soil; add organic matter before planting
HARDINESS Frost hardy at temperatures down to -5°C/23°F; needs winter protection in colder areas
PROBLEMS None
CARE Dry mulch in winter to ensure a good flower display; cut back faded foliage in late winter or early spring
PROPAGATION Sow ripe seed in a cold frame in autumn; division before growth in spring

Cynara cardunculus ♟
Cardoon

⬆ 1.5m/5ft ↔ 1.2m/4ft   **EASY**

This architectural whopper of a perennial, from Morocco and the Mediterranean, is not for the faint-hearted, but it is really exciting. It has strong, sturdy silver stems with large, deeply divided, grey-silver leaves up to 50cm/20in long. The large, slightly sweet-scented, thistle-like purple flowers produced in summer are a magnet for bees and butterflies and look lovely touched with frost. Give it plenty of room: it will need it.

BEST USES Makes a striking focal point in a gravel garden, sunny border or coastal garden; a good choice for large wildflower gardens

DROUGHT TOLERANCE Excellent, once established
FLOWERS June to September
SCENTED Slightly scented flowers
ASPECT South, west or east facing, in a sheltered position; full sun
SOIL Any fertile, well-drained soil
HARDINESS Fully hardy/borderline at temperatures down to -15°C/5°F; may need winter protection in colder areas
PROBLEMS Aphids, slugs and snails; botrytis (grey mould)
CARE Remove the flower buds to keep foliage more silvery; tidy gardeners can cut it back to ground level in winter; apply a dry mulch in winter
PROPAGATION Sow seed in a cold frame in spring; division in spring; root cuttings in winter

GREENFINGER TIP *This plant needs brawny supports: thick telephone cable will be man enough for the job*

Deutzia × kalmiiflora
Kalmia deutzia

⬆ 1.5m/5ft ⬌ 1.5m/5ft   **EASY**

This graceful arching deciduous shrub, of Asian origin, has small, oval, green leaves and masses of gorgeously fragrant clusters of white flushed pink star-shaped flowers in summer, with paler hues inside. This plant is tolerant of just about any garden soil and its charming habit is a real boost to the flower border. Give it plenty of space, so you can stand back and admire its simple beauty and sweet fragrance.

BEST USES Ideal in the summer border; also good for cottage and wildflower gardens as pollinating insects enjoy it; works well in coastal gardens

DROUGHT TOLERANCE Excellent, once established

FLOWERS June to July

SCENTED Highly fragrant flowers

ASPECT Any, in a sheltered or exposed position

SOIL Any fertile, well-drained soil

HARDINESS Fully hardy at temperatures down to -15°C/5°F; needs no winter protection

PROBLEMS None

CARE When shaping a juvenile plant, remove young growth to encourage a pleasing shape; in mature plants, remove any damaged or crossing stems to ground level; trim after flowering

PROPAGATION Sow seed in a cold frame in autumn; softwood or semi-ripe cuttings in late spring to mid-summer; hardwood cuttings in autumn

Digitalis purpurea
Common foxglove

⬆ 2m/6ft ⬌ 60cm/24in   **EASY**

Foxgloves take some beating. They are one of nature's marvels, with their stately spires of tubular flowers and intricate throat markings. This is our native, common biennial variety (making leaves the first year and flowers the next), with slightly hairy, dark green, pointed leaves in a circular formation around the stem and a solitary ascending spike of graduated pinky purple flowers with brown-maroon spotting inside (the spots act like road maps, guiding bees to the nectar). *D.p.* f. *albiflora* is a lovely white variety.

BEST USES Adds marvellous vertical accents to any flower border; will do well for the wildflower or cottage garden; very picturesque in a lightly shaded woodland garden

DROUGHT TOLERANCE Excellent, once established

FLOWERS June

SCENTED No

ASPECT Any, in a sheltered or exposed position; full sun to partial shade

SOIL Any fertile, humus-rich, well-drained soil; add organic matter before planting

HARDINESS Fully hardy at temperatures down to -15°C/5°F; needs no winter protection

PROBLEMS Powdery mildew

CARE Cut down the dead flower spikes to prevent self-seeding, if desired

PROPAGATION Self-seeds freely; surface-sow seed at 10°C/50°F or scatter in situ in late spring

Echinacea purpurea
Cone flower

⬆ 1.5m/5ft ↔ 45cm/18in   **EASY**

This underrated upright herbaceous perennial from the United States is a member of the aster family. It is as useful for its toughness as its late summer appearance in the border, with large (up to 12cm/5in across) daisy-like flowers of pinky purple that droop slightly from very prominent golden brown cone-shaped centres. It has oval leaves with pointed tips. *E.* 'Art's Pride' has orange–red flowers and *E. purpurea* 'Ruby Giant' 🏅 has beautiful dark wine-coloured petals.

BEST USES Suits prairie-style planting and naturalises well with swaying grasses; also good for wildlife and cottage gardens, as bees love it

DROUGHT TOLERANCE Good, once established
FLOWERS July to September
SCENTED No
ASPECT South, west or east facing, in a sheltered or exposed position; full sun to partial shade
SOIL Any fertile, humus-rich, well-drained soil; add organic matter before planting
HARDINESS Fully hardy at temperatures down to -15°C/5°F; needs no winter protection
PROBLEMS None
CARE Remove faded flower stems and cut back to ground level in spring; mulch in spring and after dividing
PROPAGATION Sow seed at 13°C/55°F in spring; division in spring or autumn (after flowering); root cuttings in late autumn to winter

Eremurus stenophyllus 🏅
Foxtail lily

⬆ 1.5m/5ft ↔ 60cm/24in   **EASY**

An eye-catching herbaceous perennial from central Asia. It has tuberous roots and towering, but stout, bright yellow flower spikes some 30cm/12in long that flower successively from the bottom up and fade to soft orange. The grey-green strappy leaves are at the base of the tall, straight, smooth stems. A fleeting glory of a plant, drought tolerant once its roots are down, it flowers for barely a month in summer, but if you have the space, why not spoil yourself?

BEST USES Looks stunning in a large border amongst tall grasses, and adds grandeur to a more traditional flower border; also useful in a wildlife garden, attracting pollinating insects

DROUGHT TOLERANCE Good, once established
FLOWERS June to July
SCENTED No
ASPECT South, west or east facing, in a sheltered position with protection from cold winds; full sun
SOIL Fertile, well-drained loam or sandy soil; add organic matter before planting
HARDINESS Fully hardy at temperatures down to -15°C/5°F; needs no winter protection
PROBLEMS None
CARE May need staking; mulch with leafmould or garden compost in autumn; in winter, dry mulch the crowns to protect early emerging foliage from frosts
PROPAGATION Sow seed in a cold frame in autumn; division after flowering in summer or early autumn

Foeniculum vulgare 'Purpureum'
Bronze fennel

⬆ 2m/6ft ⬌ 60cm/24in   **EASY**

This tall, feathery-leaved perennial from coastal
Mediterranean regions has long been popular for
its herbal properties, but is increasingly used as
an architectural border plant. It has smoky,
spidery, flat flower heads of small yellow flowers
in summer, with young filigreed foliage that is
purple to bronze, greying with age. The foliage
smells of aniseed when bruised.

BEST USES Amazing in the flower border,
especially with golden or acid-coloured foliage
plants; makes a stunning partner for large vivid
yellow, red or orange-flowered plants in a 'hot'
border; perfectly at home in the cottage garden;
coastal gardeners will appreciate its ferny grace

DROUGHT TOLERANCE Excellent, once established

FLOWERS August, but grown mainly for foliage

SCENTED Aromatic leaves

ASPECT South or west facing, in a sheltered position; full
sun to partial shade

SOIL Any fertile, moist, well-drained soil; add organic
matter before planting

HARDINESS Fully hardy at temperatures down to
-15°C/5°F; needs no winter protection

PROBLEMS Slugs and snails may nibble juvenile foliage

CARE Divide overcrowded clumps in late spring; cut
back to ground level in autumn

PROPAGATION Sow seed at 15–21°C/59–70°F in
spring or in situ in late spring; division in autumn every
three years

Gypsophila paniculata
Baby's breath

⬆ 1.2m/4ft ⬌ 1.2m/4ft   **EASY**

This originates in the dry hills of the
Mediterranean, but I am sure you have bought
this delightful froth from the florist at some time.
It is every bit as pretty as a herbaceous perennial
in the traditional flower border. The leaves are
small, narrow and glaucous, and hardly visible
when the plant is producing its airy clouds of
tiny white flowers in summer. The flowers will
brown in extended drought, and the plant can
be short-lived, but is easy to raise from seed.
G. 'Rosenschleier' ♀ has pale pink flowers.

BEST USES Ideal as a cottage garden or border
plant; looks beautiful grown at the feet of taller shrub
roses, as it hides their bare lower limbs and offers a
marvellous misty effect

DROUGHT TOLERANCE Good, once established

FLOWERS June to August

SCENTED No

ASPECT South, west or east facing, in a sheltered or
exposed position; full sun

SOIL Any fertile, sharp-draining soil; can be short-lived in
heavy clay

HARDINESS Fully hardy at temperatures down to
-15°C/5°F; needs no winter protection

PROBLEMS None, though can be prone to stem rot

CARE May need staking; cut back after flowering; avoid
moving as, being tap-rooted, it will die

PROPAGATION Sow seed in situ in early to
mid-spring

Melianthus major ♈
Honey bush

⬆ 2.5m/8ft ⬌ 90cm/3ft   **EASY**

This stunning, architectural, mounded evergreen shrub from South Africa has tough, hollow branching stems and the most elegant, deeply serrated grey-green foliage. It will often produce slightly arching racemes of tubular, brown-red maroon flowers (that spill nectar on the leaves), held on attractive drooping purple stems. One of the most beautiful foliage plants around: 10/10.

BEST USES The beautiful foliage is best displayed in a Mediterranean border or gravel garden but will do equally well as a specimen plant in a large container or in a traditional border

DROUGHT TOLERANCE Good, once established
FLOWERS May to July, but grown mainly for foliage
SCENTED No
ASPECT South facing, in a sheltered position protected from cold winds; full sun
SOIL Any fertile, humus-rich, well-drained soil; add organic matter before planting; dislikes sitting in winter wet
HARDINESS Half hardy at temperatures down to 0°C/32°F; needs winter protection in cold areas
PROBLEMS None
CARE Cut to ground level in late winter or early spring; protect the crown with a dry winter mulch
PROPAGATION Sow seed at 13–18°C/55–64°F in spring; basal softwood cuttings in spring

••
GREENFINGER TIP *This survives quite hard winters with a good mulch on a sandy bank: I suspect mature plants are more hardy than young ones*

Olea europaea
Olive

⬆ 3m/10ft ⬌ 3m/10ft   **MEDIUM**

As our climate grows milder, this evergreen tree is now popular, normally grown as a half-standard container specimen or a small shrub in a sheltered garden, but it will never achieve the bulk and height of the Mediterranean originals. It has small, tough, silvery grey-green leaves and fragrant white flowers in summer, followed by rounded green fruits (which ripen to black edible olives in hot climates).

BEST USES A lovely focal point against a warm sheltered wall in a Mediterranean garden; grow in containers as a half-standard in a sheltered spot; offers fantastic evergreen form in a sheltered winter garden

DROUGHT TOLERANCE Excellent, once established
FLOWERS June to August, but grown mainly for foliage
SCENTED Lightly scented flowers
ASPECT South or west facing, in a sheltered position with protection from cold winds; full sun to partial shade
SOIL Any fertile, well-drained soil
HARDINESS Frost hardy at temperatures down to -5°C/23°F; needs winter protection in all but the mildest areas
PROBLEMS Vine weevil if grown in containers
CARE Remove dead or damaged wood in spring
PROPAGATION Sow seed at 13–15°C/55–59°F in spring; semi-ripe cuttings in summer from two-year-old wood

Perovskia 'Blue Spire' 🎖
Russian sage

⬆ 1.2m/4ft ⬌ 90cm/3ft **EASY**

Russian sage is a sure-fire winner and an ever popular, low-maintenance, deciduous sub-shrub, perhaps because of its misty romantic appearance in the late summer border. It has a fairly erect habit and small, soft, narrow, finely cut, aromatic grey-green leaves. The stiff stems are smothered in massed spires of tiny violet blue flowers in summer and take on an alluring skeletal beauty in autumn, so don't be tempted to cut them back too soon.

BEST USES Incredibly attractive in a traditional or Mediterranean border; also ideal for naturalised gardens (insects love it) or coastal gardens

DROUGHT TOLERANCE Excellent, once established

FLOWERS July to September

SCENTED Aromatic leaves

ASPECT South, west or east facing, in a sheltered or exposed position; full sun

SOIL Any moderately fertile, dry, well-drained soil, especially chalk; avoid waterlogged soil

HARDINESS Fully hardy at temperatures down to -15°C/5°F; needs no winter protection

PROBLEMS None

CARE Cut the flowering stems back hard to the base in early spring

PROPAGATION Softwood cuttings in spring; semi-ripe cuttings in summer

••

GREENFINGER TIP *Don't attempt to plant this in anything but sharply draining soil: it will simply sulk for a very short time and then die on you!*

Romneya coulteri 🎖
Tree poppy/California tree poppy

⬆ 1.2m/4ft ⬌ Indefinite **MEDIUM**

This suckering perennial is mercurial in nature, but if you can get it to behave it really is a lovely plant, with deeply jagged grey-green leaves and large, fragrant, tissue-paper flowers of pure white with golden centres. The flowers resemble a slightly wrinkled fried egg! Some people struggle to get this plant established, but for others it leaps around the garden with great alacrity. With its cheerful breezy manner, it is simply delightful.

BEST USES Best on a warm sheltered wall in the herbaceous border; also suits a Mediterranean or cottage garden

DROUGHT TOLERANCE Excellent, once established

FLOWERS July to October

SCENTED Lightly fragrant flowers

ASPECT South or west facing, in a sheltered position with protection from cold winds; full sun

SOIL Any fertile, humus-rich, well-drained soil; add organic matter before planting

HARDINESS Frost hardy at temperatures down to -5°C/23°F; needs winter protection in colder areas

PROBLEMS Caterpillars; powdery mildew and *Verticillium* wilt; can be temperamental in establishing

CARE Cut back hard to main framework in spring; mulch in winter; may be damaged by frosts, but will recover

PROPAGATION Sow seed soaked in alcohol in modules (to avoid root disturbance) in a cold frame in autumn; root cuttings (placed horizontally rather than vertically) in winter

Rosa Golden Celebration ♀
(also known as 'Ausgold')

⬆ 1.2m/4ft ⬌ 1.2m/4ft **EASY**

This lovely English shrub rose can also be grown as a short climber. Some roses are a joy to behold and this is one of them. With light green leaves and a rounded habit, it's the generous blooms that are a knockout. Tight, yellow buds suffused with rosy orange open to unusually large, cupped, warm yellow flowers with a 'proper' tea-scented fragrance, held on slightly arching stems. It is repeat-flowering.

BEST USES Grow as a small climber over a pergola, low wall or fence, or as a shrub in the mixed border; does well in containers and is excellent as a cut flower

DROUGHT TOLERANCE Excellent, once established

FLOWERS June to September

SCENTED Yes

ASPECT South or west facing, in a sheltered position; full sun

SOIL Any fertile, well-drained soil; add organic matter before planting

HARDINESS Fully hardy at temperatures down to -15°C/5°F; needs no winter protection

PROBLEMS Greenfly; has good disease resistance

CARE Mulch annually with organic matter; deadhead after flowering and cut back flowering stems, leaving 2–3 buds on this season's new growth; prune back by a third in winter or when growth starts in spring in colder regions

PROPAGATION Hardwood cuttings in autumn

Rosa 'Roseraie de l'Haÿ' ♀

⬆ 1.8m/6ft ⬌ 1.5m/5ft **EASY**

Of Asian heritage, and widely regarded as one of the best of the rugosas, this fairly fast-growing, deciduous, sturdy bushy shrub has apple green leaves and very prickly strong stems. Its charm lies in the profusion of clusters of almond-perfumed, loose, double purple crimson flowers that have attractive cream stamens and are produced in joyful abundance from summer into early autumn, with the odd bloom in October.

BEST USES Makes a lovely informal flowering hedge, which is a good animal barrier and ideal for coastal gardens; can be grown as a single shrub but has more visual impact in multiples

DROUGHT TOLERANCE Excellent, once established

FLOWERS July to September

SCENTED Yes

ASPECT Any, in a sheltered or exposed position; full sun to partial shade

SOIL Any fertile, well-drained soil; add organic matter before planting

HARDINESS Fully hardy at temperatures down to -15°C/5°F; needs no winter protection

PROBLEMS None

CARE Needs no pruning; trim hedges in winter after flowering to promote bushy growth

PROPAGATION Hardwood cuttings in autumn

GREENFINGER TIP *When planting this as a hedge, dig a furrow and line with organic matter; plant, then backfill with a good compost; firm in well and water until established*

Verbena bonariensis 🎖
Purple topped/Tall verbena

⬆ 2m/6ft ↔ 50cm/20in   **EASY**

There are plants and there are fabulous plants, and this is one of the latter. It is a tall herbaceous perennial, with very erect, skeletal, self-supporting olive-grey stems, sparse leaf coverage and the most heavenly heads of lilac purple that lift themselves graciously above the clamour of lower-growing perennials and are madly attractive to butterflies and bees. It self-seeds with great abandon, but is never invasive or a nuisance. Deserves 10/10 for attracting wildlife and for being so gorgeously congenial. Frankly, you cannot live without it.

BEST USES Try this in the middle or front of any border; looks absolutely tip-top with gaura, gypsophila and both common or purple angelica

DROUGHT TOLERANCE Excellent, once established
FLOWERS June to September
SCENTED No
ASPECT South or west facing, in a sheltered position with protection from cold winds; full sun
SOIL Any fertile, well-drained soil; will struggle in very waterlogged soil
HARDINESS Fully hardy at temperatures down to -15°C/5°F; needs no winter protection
PROBLEMS None
CARE Cut back old stems in spring, once new basal growth begins to emerge; strong winds can force this plant sideways, so it may need staking
PROPAGATION Self-seeds freely; sow seed at 21°C/70°F in spring; division in spring

Yucca gloriosa 🎖
Spanish dagger

⬆ 2m/6ft ↔ 2m/6ft   **EASY**

This large, woody, spiky evergreen shrub is statuesque to say the least, so will require plenty of space. It has long, sharply pointed, very rigid linear bladed leaves that are bluish-green but mature to deep green. Very large spiky sprays of whitish-cream egg-shaped flowers are produced in isolated displays in late summer and often into early autumn. The exotic-looking foliage is striking (and injurious!) all year round.

BEST USES Perfect for a focal point in a gravel or Mediterranean garden; will do extremely well for coastal gardens; can also be grown against a warm, sheltered wall in a cottage or informal garden

DROUGHT TOLERANCE Excellent, once established
FLOWERS August to September
SCENTED No
ASPECT South or west facing, in a sheltered position; full sun
SOIL Any well-drained soil
HARDINESS Frost hardy at temperatures down to -5°C/23°F; may need winter protection
PROBLEMS Aphids; leaf spot
CARE Trim lightly after flowering and cut out any dead or damaged material
PROPAGATION Sow pre-soaked seed with bottom heat at 15°C/59°F in spring; pot up rooted suckers in spring

GREENFINGER TIP *The leaves are needle sharp, so take care when handling and don't plant where children play*

AUTUMN

There are some really cracking drought-tolerant plants around that provide spectacular autumn interest. If the summer has been a long, dry one, autumn is the time when you will appreciate even more the advantages of having given room to plants in the garden that can hold their own in times of water deficiency. The disappointment of those lush herbaceous perennials that fried and died over the summer will soon be forgotten as you enjoy the many splendours of the autumn drought-lovers.

Carex comans 'Frosted Curls'

⬆ 20cm/8in ⬅➡ 60cm/24in   **EASY**

Sedges normally like a moist environment, but this compact evergreen perennial from New Zealand will tolerate a wide variety of conditions except sitting in winter wet. It has pale silvery white, arching, linear foliage with curling stem tips that resembles a well-groomed upturned mop, and small, insignificant brown flower spikes in summer.

BEST USES Makes great ground cover; useful for naturalised planting schemes in prairie style or cottage gardens; will do well in containers in city gardens and patios

DROUGHT TOLERANCE Good, once established

FLOWERS June to October, but grown mainly for foliage

SCENTED No

ASPECT South, west or east facing, in a sheltered position; full sun

SOIL Any well-drained, fertile soil; will not thrive in waterlogged soil

HARDINESS Frost hardy at temperatures down to -5°C/23°F; needs winter protection in colder areas

PROBLEMS Aphids

CARE Comb out dead foliage with your fingers in spring and summer

PROPAGATION Sow seed at 15°C/59°F in spring; division in spring; pot up rooted shoots in early summer

Erigeron karvinskianus ♔
Mexican fleabane/Spanish daisy

⬆ 20cm/8in ⬌ 60cm/24in   EASY

This small, mat-forming rhizomatous perennial from Mexico and Panama bears an extravagant profusion of small daisy-like flowers that have white petals flushed with pink, and cheerful yellow centres. The mid-green leaves are small and ferny. It is one of the handiest plants I know, and the long flowering period is a real bonus: it begins in early summer and can go through into November in mild areas. This is one plant I could not do without. It's not posh, but boy does it work!

BEST USES Spectacular tumbling down walls and covering dry banks; will pop up between paving and in drystone walls; does well underplanting larger specimens in containers; makes effective, pretty ground cover in a gravelled area

DROUGHT TOLERANCE Excellent, once established

FLOWERS June to October

SCENTED No

ASPECT South, west or east facing, in a sheltered position; full sun

SOIL Any well-drained, fertile soil

HARDINESS Fully hardy at temperatures down to -15°C/5°F; needs no winter protection

PROBLEMS None

CARE Cut back after flowering to prevent excessive self-seeding; trim lightly in spring (or hard if unruly) to prevent it getting straggly; divide overcrowded clumps in spring

PROPAGATION Self-seeds freely; sow seed at 15°C/59°F in spring; division in spring

Geranium clarkei 'Kashmir White' ♔

⬆ 45cm/18in ⬌ 90cm/3ft   EASY

A spreading herbaceous hardy perennial, originally from India (hence 'Kashmir'), with finely cut, divided mid-green leaves, making large, pleasing mounds of foliage. The simple, gappy, white-cupped flowers with pale pink veining are held delicately above the foliage. The foliage is an appealing feature in itself from late spring until autumn, and this is also well worth growing for its longer than average flowering period.

BEST USES A mainstay of the cottage garden and ideal as ground cover or underplanting in a shrub border; planted repeatedly through a herbaceous border, it supplies subtle rhythm and continuity

DROUGHT TOLERANCE Good, once established

FLOWERS July to October

SCENTED No

ASPECT Any, in a sheltered or exposed position; full sun to partial shade

SOIL Any well-drained, fertile soil; add organic matter before planting; will not thrive in waterlogged soil

HARDINESS Fully hardy at temperatures down to -15°C/5°F; needs no winter protection

PROBLEMS Capsid bug and vine weevil; powdery mildew

CARE Cut back to ground level in early spring; divide overcrowded clumps in early spring or late summer to autumn; remove old flowering stems in summer to encourage repeat flowering

PROPAGATION Sow seed at 15°C/59°F in spring; division in spring or autumn

Hypericum calycinum
Rose of Sharon/St John's wort

⬆ 60cm/24in ⬌ Indefinite   **EASY**

This evergreen shrub from Turkey seems to be epidemic at the moment – I certainly see it wherever I go. I suppose it is a dependable, useful plant that will grow almost anywhere. It has dark green, oval leaves with paler undersides and gaudy buttercup yellow flowers with pronounced yellow stamens. It spreads by runners and is difficult to get rid of once established. I loathe it, but can see why it is so popular. Not in my garden however!

BEST USES Will cover awkward banks, slopes and verges and even survive well under hedges and small trees

DROUGHT TOLERANCE Good, once established

FLOWERS June to October

SCENTED No

ASPECT Any, in a sheltered or exposed position; full sun to partial shade

SOIL Any well-drained, fertile soil

HARDINESS Fully hardy at temperatures down to -15°C/5°F; needs no winter protection

PROBLEMS Rust

CARE Cut down to ground level in early spring to reduce the risk of rust (however, I have never yet met a plant that didn't suffer from it)

PROPAGATION Semi-ripe cuttings in summer

Liriope muscari ♈
Blue lilyturf

⬆ 30cm/12in ⬌ 20cm/8in   **EASY**

A reliable, stout, clump-forming tuberous perennial from Japan and China that copes well with dry shade once it has had time to establish. They have strappy, evergreen dark green leaves and produce short, straight stems of intense violet-coloured flower spikes in late summer to early autumn.

BEST USES Ideal for ground cover in shaded woodland areas and awkward slopes; equally pleasing in containers in a city garden or at the front of beds in the cottage garden

DROUGHT TOLERANCE Excellent, once established

FLOWERS September to October

SCENTED No

ASPECT Any, in a sheltered position; full sun to full shade

SOIL Any fertile, moist, well-drained soil; add organic matter before planting

HARDINESS Fully hardy at temperatures down to -15°C/5°F; needs no winter protection

PROBLEMS Slugs

CARE Cut away old leaves in spring to encourage fresh growth

PROPAGATION Sow seed in pots outdoors in spring; division in spring

Nerine bowdenii ♀

⬆ 45cm/18in ⬌ 45cm/18in   **MEDIUM**

This bulbous perennial from South Africa is a mainstay of the autumn flower border for its unexpectedly spring-like flowers. It has fresh green, strappy leaves that appear after the large, musky-scented, candy pink trumpet-shaped flowers open, looking startlingly exotic so late in the year. They love long hot summers, with their feet baking hot, their heads in the sun, crowded in with other convivial nerines for company, so choose a planting site with this in mind and they will quickly establish.

BEST USES Do well undisturbed in containers; the colour effect is fantastic planted in blocks in the border or gravel garden; excellent as cut flowers; naturalise well with short ornamental grasses

DROUGHT TOLERANCE Excellent, once established

FLOWERS September to November

SCENTED Lightly scented flowers

ASPECT South, west or east facing, in a sheltered position; full sun

SOIL Any reasonable, well-drained soil; does well in poor, dry soils

HARDINESS Fully hardy at temperatures down to -15°C/5°F; needs no winter protection

PROBLEMS Slugs

CARE Plant bulbs in autumn or early winter and apply an organic mulch to help them establish

PROPAGATION Sow ripe seed immediately at 10–13°C/50–55°F; division after flowering (you will need a sturdy fork!)

Ophiopogon planiscapus 'Nigrescens' ♀
Black lilyturf

⬆ 30cm/12in ⬌ 20cm/8in   **EASY**

This small, clumping, evergreen perennial with Asian origins became all the rage when garden makeover programmes introduced it to our television screens, and no wonder. It is grown mainly for its spidery, grass-like blades of purply charcoal black but also has very pretty sprays of grey-purple flowers, followed by tiny black berries. Very urban. Very chic and still very 'now', with the unabated trend for naturalised planting.

BEST USES Plant in multiples for the best effect; works well in containers or at the front of a hot, dry, sunny border, where it is not overshadowed by surrounding plants and looks remarkable with *Alchemilla mollis*; good for coastal conditions

DROUGHT TOLERANCE Excellent, once established

FLOWERS June to August, but grown mainly for foliage

SCENTED No

ASPECT South, west or east facing, in a sheltered or exposed position; full sun to partial shade

SOIL Almost any fertile, well-drained soil; a slightly acid soil sees the colour at its best

HARDINESS Fully hardy at temperatures down to -15°C/5°F; needs no winter protection

PROBLEMS Slugs

CARE Remove dead or diseased leaves, as necessary

PROPAGATION Division in spring

GREENFINGER TIP *This may not clump up as quickly as one might like, but it will colonise happily if left alone. Patient gardeners will prevail!*

Pennisetum orientale 'Karley Rose'
Fountain/Foxtail grass

⬆ 60cm/24in ↔ 75cm/30in   EASY

Pennisetums are clump-forming perennial grasses that look remarkably natural in a garden setting. This compact variety has fine, bright green, gracefully tapered leaves and forms tidy rounded hummocks. It produces masses of soft, feathery plumes of creamy buff colour that mature to pale wine hues from summer to autumn. Planted in large swathes or graceful arcs, their gentle movement when stirred by soft winds is simply magical.

BEST USES Ideal for the wildflower, wildlife or prairie-style garden but looks equally well grown in pots in a chic urban setting or roof terrace

DROUGHT TOLERANCE Excellent, once established

FLOWERS July to September, but grown mainly for foliage

SCENTED No

ASPECT South, west or east facing, in a sheltered position with protection from cold winds; full sun

SOIL Any fertile, well-drained soil

HARDINESS Fully hardy at temperatures down to -15°C/5°F; needs no winter protection

PROBLEMS None

CARE Comb out dead or tatty growth with your fingers; cut back hard in late winter or early spring as the new shoots appear

PROPAGATION Self-seeds freely; sow seed at 13–18°C/50–55°F in spring; division in spring

Sedum 'Ruby Glow' ⚇
Stonecrop

⬆ 25cm/10in ↔ 45cm/18in   EASY

This hardy, low-growing, deciduous perennial has fleshy pale green stems, succulent toothed leaves and produces myriad tiny star-shaped flowers that form large, dense, heavy flower heads that are pale green in summer and open to a rich claret in early autumn. Its interesting architectural shape makes it attractive through the early summer, and the changing colour of the flower heads in autumn is always intriguing.

BEST USES Ideal for the wildflower or cottage garden as bees and butterflies love this plant

DROUGHT TOLERANCE Excellent, once established

FLOWERS July to September; flowers colour up in autumn

SCENTED No

ASPECT South or west facing, in a sheltered or exposed position; full sun to partial shade

SOIL Any fertile, well-drained soil

HARDINESS Fully hardy at temperatures down to -15°C/5°F; needs no winter protection

PROBLEMS None, though slugs, snails and vine weevil are a potential nuisance

CARE Cut back to the base in early spring; divide every three years to avoid an unsightly central gap

PROPAGATION Sow seed at 13–16°C/55–61°F in autumn; division in spring or late autumn

GREENFINGER TIP *The flower heads are liable to collapse outwards, so stake early in the year*

Sternbergia lutea
Winter daffodil

⬆ 15cm/6in ⬅➡ 8cm/3in   EASY

This bulbous perennial is very free-flowering once established, with narrow, linear dark green leaves and short stems producing goblets of single golden yellow flowers, rather like crocus, from early autumn. They remain dormant from spring, but need a dry, hot summer to produce flowers, which can then withstand almost any autumn temperatures. An excellent plant for autumn colour.

BEST USES Ideal for a very hot, dry, sunny border, rock garden, gravelled area or containers; will survive in beds and borders, but the soil needs to be very well drained and baking hot in summer

DROUGHT TOLERANCE Excellent, once established

FLOWERS September to October

SCENTED No

ASPECT South, west or east facing, in a sheltered position; full sun to partial shade

SOIL Any fertile, sharply drained soil

HARDINESS Fully hardy at temperatures down to -15°C/5°F; needs no winter protection

PROBLEMS Narcissus bulb fly and narcissus eelworm

CARE Divide large colonies in spring if flowering is poor or blind, to encourage flowering; divide offsets in summer to early autumn

PROPAGATION Plant bulbs in summer

GREENFINGER TIP *They seem to clump up better in lighter soils*

Zauschneria californica 'Dublin' ♀
Californian fuschia

⬆ 30cm/12in ⬅➡ 50cm/20in   EASY

This is a very exotic-looking, rhizomatous, low-growing deciduous sub-shrub, normally found in dry rocky places in North America, where it is pollinated by hummingbirds. It has small, linear grey-green leaves and bears very showy bright red flowers that are slightly drooping and trumpet-shaped (perfect for the birds' long thin beaks) on slender, plum-coloured stems. It delivers a blast of exotic colour in autumn, but needs a hot, dry baking spot to see it at its best.

BEST USES Beds and borders will benefit from the bold colour; will do well in the Mediterranean or gravel garden, and in rock gardens or troughs; grow it in containers if you are on heavy soil

DROUGHT TOLERANCE Excellent, once established

FLOWERS July to November

SCENTED No

ASPECT South or west facing, in a sheltered position with protection from cold winds; full sun

SOIL Any fertile, well-drained soil, especially poor, stony soils

HARDINESS Fully hardy/borderline at temperatures down to -15°C/5°F; may need winter protection in colder areas

PROBLEMS Slugs and snails like young growth

CARE Trim to about 10cm/4in from the base after all danger of frosts has passed

PROPAGATION Sow seed in a cold frame in spring (bottom heat improves germination); basal stem cuttings in late spring

Agastache 'Black Adder'
Hyssop

⬆ 90cm/3ft ⬌ 40cm/16in   **EASY**

Invaluable for their long flowering period, hyssops are native to dry hillsides of the USA and Mexico, which accounts for their drought tolerance. This busy, herbaceous hardy perennial has liquorice-scented, lance-shaped, fresh green leaves and brushes of smoky coloured spikes made up of myriad small, tubular, deep rich purply blue flowers that are irresistible to butterflies and bees. The foliage is attractive from spring onwards and the plant is definitely worth growing for its vibrant autumn colour and mass of flowers over a long period.

BEST USES Perfect for adding height and colour to the autumn border; also provides marvellous ground cover; ideal for encouraging pollinating insects so a good subject for the wildlife garden

DROUGHT TOLERANCE Good, once established
FLOWERS July to October (to the first frosts)
SCENTED Aromatic leaves
ASPECT South or west facing, in a sheltered or exposed position; full sun
SOIL Any fertile, well-drained soil; add organic matter before planting; dislikes sitting in winter wet
HARDINESS Frost hardy at temperatures down to -5°C/23°F; needs winter protection in colder areas
PROBLEMS Powdery mildew
CARE Cut back hard in early spring
PROPAGATION Sow seed at 15°C/60°F in spring; division in spring; semi-ripe cuttings in summer or autumn

Aster turbinellus 🎖
Michaelmas daisy

⬆ 90cm/3ft ⬌ 60cm/24in   **EASY**

Asters are originally from southern Europe and the prairies of North America, which gives an indication of the diverse habitat they enjoy. This variety is a clump-forming herbaceous perennial with dark green lance-shaped leaves and upright, slender purple-brown stems bearing light sprays of blue-violet daisy-like flowers with golden eyes borne in fair abundance in autumn. They are well worth growing in pots to plunge into bare gaps in the late summer border.

BEST USES Ideal for the cottage garden, but also useful for prairie style planting; attractive to late bees in the wildlife garden; lasts well as a cut flower

DROUGHT TOLERANCE Good, once established
FLOWERS July to October
SCENTED No
ASPECT South, west and east facing, in a sheltered or exposed position; full sun
SOIL Any fertile, well-drained soil; add organic matter before planting
HARDINESS Fully hardy at temperatures down to -15°C/5°F; needs no winter protection
PROBLEMS Aphids, slugs and snails; botrytis (grey mould)
CARE Cut back to ground level in autumn after flowering
PROPAGATION Sow seed at 15°C/59°F or in a cold frame in spring; division in spring

Centranthus ruber
Red valerian

⬆ 90cm/3ft ↔ 90cm/3ft   **EASY**

Originating from the Mediterranean to Turkey, this woody-based, clump-forming perennial is widely regarded as a pernicious weed. It is fairly stiffly erect in habit and has slightly fleshy, lance-shaped leaves, producing clusters of tiny, reddish pinky flowers reliably from summer well into autumn. It self-seeds freely and you will need to keep pulling it up if it wanders where it shouldn't. The white version *C.r.* 'Albus' is marginally less drought tolerant until established, but is also less invasive and, for my money, a better choice, as I can't bear the pink variety!

BEST USES Ideal for the natural cottage garden, with its informal habit; good for wildflower gardens as it attracts insects; useful in coastal areas as it is extremely tolerant of salty winds

DROUGHT TOLERANCE Excellent, once established

FLOWERS July to October

SCENTED No

ASPECT South or west facing, in a sheltered or exposed position; full sun

SOIL Any poor to fertile, well-drained soil

HARDINESS Fully hardy at temperatures down to -15°C/5°F; needs no winter protection

PROBLEMS None

CARE Deadhead through summer to limit self-seeding; cut down stalks in autumn

PROPAGATION Self-seeds freely; sow seed in a cold frame in spring; division in spring

Ceratostigma willmottianum 🎖
Hardy plumbago/Chinese plumbago

⬆ 90cm/3ft ↔ 1.5m/5ft   **EASY**

From China again (so many beautiful plants come from there), this really useful, small, deciduous shrub has slightly hairy mid-green leaves that are flushed with purple as autumn approaches. It is smothered in simple flowers of the clearest blue for a long period, sometimes right up to the first frosts, and is an indispensable late performer for the autumn flower border. Get yourself one.

BEST USES I am trying to think of a garden setting that wouldn't suit this plant . . . and I can't

DROUGHT TOLERANCE Good, once established

FLOWERS June to October

SCENTED No

ASPECT South, west or east facing, in a sheltered or exposed position; full sun to partial shade

SOIL Any fertile, well-drained soil

HARDINESS Fully hardy/borderline at temperatures down to -15°C/5°F; may need winter protection in very cold areas

PROBLEMS Powdery mildew; the top-growth may be damaged by frosts, but renewed growth will appear in spring

CARE Cut back the flowering stems after flowering to about 2.5cm/1in of the old woody growth

PROPAGATION Softwood cuttings in early summer; hardwood cuttings in winter

Cotoneaster horizontalis 🏅
Wall spray

⬆ 90cm/3ft ⬅➡ 1.5m/5ft   **EASY**

The many virtues of this small deciduous shrub from China are often overlooked as it is so common, but it is an incredibly versatile, low-maintenance plant. It has small, round, dark green glossy leaves that mature to red orange in autumn, with bright red berries. Though grown mainly for its autumn berry display, it bears small, pretty pinkish flowers in spring. The flat, fishbone-patterned stems are very eye-catching even in deepest winter, and can be easily trained up a wall. This is one of those rare shrubs that look good all year round.

BEST USES Ideal for growing around porches and to frame doorways or against low brick walls; the berries will encourage birds into the garden

DROUGHT TOLERANCE Excellent, once established

FLOWERS May, but grown mainly for autumn berries

SCENTED No

ASPECT South, west or east facing, in a sheltered or exposed position; full sun to full shade

SOIL Any fertile, well-drained soil

HARDINESS Fully hardy at temperatures down to -15°C/5°F; needs no winter protection

PROBLEMS Woolly aphids; fireblight

CARE Minimal; trim in late winter or early spring, removing dead or crossed stems

PROPAGATION Layering in early spring; semi-ripe cuttings in mid-summer to autumn

Panicum virgatum 'Heavy Metal'
Switch grass

⬆ 90cm/3ft ⬅➡ 75cm/30in   **EASY**

If you like grasses, you will love this attractive, sturdy, deciduous, clumping perennial. It has fairly upright, blue metallic leaf colouring and pink, gauzy flower heads, held in soft wisps. It ripples in soft waves when stirred by a breeze and, like all the grasses, is incredibly versatile in creating naturalistic planting schemes.

BEST USES Will grow well in a pot in an urban or rooftop garden; naturalises well, so looks equally good in groups in a gravel or Mediterranean garden or mixed in with flowering perennials

DROUGHT TOLERANCE Good, once established

FLOWERS July to October, but grown mainly for foliage

SCENTED No

ASPECT South, west or east facing, in a sheltered or exposed position; full sun

SOIL Any fertile, well-drained soil

HARDINESS Fully hardy at temperatures down to -15°C/5°F; needs no winter protection

PROBLEMS None

CARE Cut back straggly and faded foliage in early spring; divide every three years or so

PROPAGATION Self-seeds freely; sow seed at a minimum of 13°C/55°F in late spring or autumn; division in spring to early summer

Agave americana ♀
Century plant

⬆ 3m/10ft ⬌ 2m/6ft   MEDIUM

This spectacular evergreen is a statuesque perennial with an exciting, prehistoric sculptural form. It has large, sharply pointed, spiny-edged, blue-green, sword-like succulent leaves: bend down too near it and it can have your eye out! Some people put corks on the ends of the leaves, but don't plant it in a garden where children play. Mature plants produce small yellowish flowers, but can take up to forty years to flower in cold regions, so you are in for a long wait.

BEST USES Wonderful as a focal point in dry gravel gardens and I espied it in a dry river bed garden in East Anglia looking ruddy marvellous

DROUGHT TOLERANCE Excellent, once established
FLOWERS Grown for foliage
SCENTED No
ASPECT South or west facing, in a sheltered position; full sun
SOIL Well-drained loamy and sandy, slightly acid soil; indoors in a container with cactus compost
HARDINESS Frost tender at temperatures down to 5°C/41°F; best grown in a container in a frost-free environment and moved outside in summer
PROBLEMS Scale insect on young growth
CARE Minimal; only re-pot if absolutely necessary as this can be perilous, don't water it in winter
PROPAGATION Sow seed at 21°C/70°F from spring to summer; better by division (for the brave) by single rosettes in spring to summer as seeds can take up to three years to get a small plant

Brugmansia aurea
Golden angels' trumpet

⬆ 5m/16ft ⬌ 2m/6ft   MEDIUM

Originally from South America, these tender evergreen shrubs (which used to be known as datura) are the ultimate in exotic elegance. This is one of the few scented varieties, the scent being stronger at night. The large-lobed green leaves are an attractive feature, but it is the large pendant, trumpet-shaped, yellow-white flowers, up to 25cm/10in long, that are so alluring. All parts of this plant are hallucinogenic and highly toxic if eaten, causing possibly fatal seizures.

BEST USES Spectacular as a specimen in a large container on a sheltered, sunny patio, filling a gap in a late summer border or in city gardens

DROUGHT TOLERANCE Good, once established
FLOWERS June to October (to the first frosts)
SCENTED Night-scented flowers
ASPECT South or west facing, in a sheltered position; full sun
SOIL Any fertile, well-drained soil
HARDINESS Frost tender at temperatures down to 7°C/45°F; best grown in a pot in a frost-free greenhouse and moved outside in summer
PROBLEMS Aphids and red spider mite if grown in a greenhouse or conservatory
CARE Deadhead regularly; cut flowered shoots to within 2–3 buds of the base in early spring; remove dead and straggly stems
PROPAGATION Sow seed at 20–25°C/69–77°F in spring; softwood or semi-ripe cuttings with bottom heat in spring or summer

Cortaderia selloana 'Sunningdale Silver' ♉ Pampas grass

⬆ 3m/10ft ⬌ 2.5m/8ft   **EASY**

This clump-forming evergreen perennial comes from the plains of New Zealand and South America, where you can imagine it thriving handsomely in wild prairie landscapes. This evergreen variety has slender, mid-green, lightly arching leaves with tall, tufted, feathery plumes of creamy-silver flower heads in late summer. Pampas grass has fallen from favour of late, but it is great value for year-round interest and splendid indeed when planted in the right place.

> **BEST USES** Beautiful as a fountain-like focal point in a hot, dry border or gravel garden or grown in pots in a Mediterranean garden

DROUGHT TOLERANCE Excellent, once established

FLOWERS August to September, but grown mainly for foliage

SCENTED No

ASPECT South, west or east facing, in a sheltered or exposed position; full sun

SOIL Any fertile, well-drained soil; add organic matter before planting

HARDINESS Fully hardy at temperatures down to -15°C/5°F; needs no winter protection after the first growing season

PROBLEMS None

CARE Comb out any dead or damaged foliage (wear gloves or your hands will be shredded) and cut tough flower stems at the base in winter or early spring with secateurs (mind out for birds nesting in the foliage)

PROPAGATION Division of large clumps in spring

Gaura lindheimeri ♉

⬆ 1.5m/5ft ⬌ 90cm/3ft   **EASY**

This graceful, gauzy clump-forming perennial from North America has lance-shaped mid-green leaves on slender, arching, trembling stems that produce spikelets of single star-shaped white flowers with spidery anthers. The flower buds are pink-tinged before opening. It dances in the breeze, with the pretty flowers as tremulous as whirling butterflies. It can be short-lived, so take cuttings as insurance, but the flowers go on and on.

> **BEST USES** Gorgeous in the cottage garden or traditional border; mixes well with grasses and naturalised planting; ideal for the wildlife garden as bees and butterflies love it

DROUGHT TOLERANCE Excellent, once established

FLOWERS June to October (to the first frosts)

SCENTED No

ASPECT South or west facing, in a sheltered position; full sun to partial shade, though flowering is reduced in shade

SOIL Any fertile, well-drained soil; add organic matter before planting

HARDINESS Fully hardy at temperatures down to -15°C/5°F; needs no winter protection

PROBLEMS Aphids and red spider mite if grown in a greenhouse or conservatory

CARE Cut back to just above ground level in early spring as new foliage emerges

PROPAGATION Self-seeds easily; sow seed in a cold frame in spring; division in spring; heeled semi-ripe cuttings in summer

Lespedeza thunbergii ♟
Bush clover

⬆ 2m/6ft ⬌ 3m/10ft   EASY

Another of the enviable offerings from China and Japan, this woody perennial (or sub-shrub) is not seen about nearly as much as I would expect but is worth seeking out for its extravagance of autumn colour. It has deeply divided, blue-green trifoliate leaves and arching stems that bear a profusion of rosy purple, pea-like flowers (it comes from the same family as the pea). It flowers like mad from summer well into autumn, so much so that the stems skim the ground under the weight. This can grow up to 1.2m/4ft in one season, so place it where it has plenty of space.

BEST USES Will do well on dry, sloping banks; also useful in the middle of borders

DROUGHT TOLERANCE Excellent, once established

FLOWERS June to October

SCENTED No

ASPECT South, west or east facing, in a sheltered or exposed position; full sun to partial shade

SOIL Any fertile, well-drained soil; add organic matter before planting

HARDINESS Fully hardy at temperatures down to -15°C/5°F; needs no winter protection

PROBLEMS None

CARE Cut back hard to main framework in spring as new growth begins; mulch annually with organic matter

PROPAGATION Greenwood cuttings in early summer; division in spring

Miscanthus sinensis 'Silberfeder' ♟

⬆ 2.5m/8ft ⬌ 1.2m/4ft   EASY

This tall, tufted perennial grass from areas across Africa to Asia adds understated elegance to any planting scheme and can be used to great softening effect. The erect, linear green leaves look gorgeous as it steadily heightens in spring, reaching its full majesty in the late summer and early autumn border when very free-flowering pale pink and beige panicles are produced, lasting well into winter. These look magnificent touched by frost. The striped foliage of *M.s.* 'Zebrinus' ♟ is also highly recommended.

BEST USES Plant in large swathes, ribboning through the traditional or Mediterranean border, and highlight by growing with taller, yellow or purple-flowering perennials; gives a contemporary feel to containers in the urban garden or roof terrace

DROUGHT TOLERANCE Excellent, once established

FLOWERS August to September

SCENTED No

ASPECT South, west or east facing, in a sheltered or exposed position; full sun

SOIL Any fertile, well-drained soil; dislikes sitting in winter wet

HARDINESS Fully hardy at temperatures down to -15°C/5°F; needs no winter protection

PROBLEMS None

CARE Cut back to ground level in spring

PROPAGATION Does not come true from seed; division of larger clumps in early to mid-spring, as new growth emerges, but is slow to re-establish

Pyracantha 'Mohave'
Firethorn

⬆ 4m/13ft ↔ 5m/16ft   **EASY**

Pyracanthas originally come from Asia and southern Europe, and this thorny, evergreen variety is an upright, bushy shrub with small glossy oval leaves, and can be successfully trained as a wall climber. It has masses of hawthorn-like white flowers in late spring, but is grown mainly for its evergreen foliage and vivid red to orange berries, which it produces in fair profusion in autumn, lasting well into winter.

> **BEST USES** Can be used as hedging, a wall climber or as a feature shrub in the mixed border, as it offers year-round interest; ideal in the wildlife garden, as the berries make it attractive to winter-foraging birds

DROUGHT TOLERANCE Excellent, once established

FLOWERS May, but grown mainly for evergreen foliage and autumn berries

SCENTED No

ASPECT South, west or east facing, in a sheltered or exposed position; full sun to partial shade

SOIL Any fertile, well-drained soil

HARDINESS Fully hardy at temperatures down to -15°C/5°F; needs no winter protection

PROBLEMS Woolly aphids; fireblight and scab

CARE Cut out dead or damaged stems in spring; trim lightly after flowering to maintain size and spread

PROPAGATION Semi-ripe cuttings in late summer to autumn; hardwood cuttings in late autumn to mid-winter

Stipa gigantea ♛
Golden oats

⬆ 2.5m/8ft ↔ 1.2m/4ft   **EASY**

One of the best grasses around, this semi-evergreen perennial from Spain and Portugal is a really lovely plant. It forms dense clumps of grey-green linear leaves, with long slender stems of silvery green ripening to tawny gold in the summer and early autumn and producing delicate, buff-coloured dangling oaty panicles. It culminates in a shimmering, diaphanous summer haze that looks good well into autumn, and the seed heads add winter interest as well: 10/10!

> **BEST USES** An absolute must for prairie-style planting; useful as a focal point in a mixed border or gravel garden; growing with lower perennials underpins its airy grace with bold splashes of flower colour; survives well in containers

DROUGHT TOLERANCE Excellent, once established

FLOWERS June to October

SCENTED No

ASPECT Any, in a sheltered or exposed position; full sun

SOIL Any fertile, well-drained soil

HARDINESS Fully hardy at temperatures down to -15°C/5°F; needs no winter protection

PROBLEMS None

CARE Cut back the flowering stalks in late winter; comb through with your fingers in early spring to remove dead or tatty growth

PROPAGATION Sow seed at 15°C/59°F in spring; division in mid-spring or early summer

WINTER

It seems odd relating drought to the winter months, when we tend to think of this as a season that embodies cold weather and rain a-plenty! However, plants that are at their best in winter have had to survive the weather all through the year to be in fine fettle for their attractive wintry displays: they may have been subject to pitiful rainfall during long, hot summers, but they still need to perform come the winter. Treat them well early in the year and they will perform admirably for you when their time comes.

Bergenia cordifolia 'Purpurea' 🎖
Elephant's ears

⬆ 60cm/24in ⬌ 75cm/30in   EASY

We have the meadows and woodlands of Asia to thank for these clump-forming evergreen rhizomatous perennials. Although they flower in spring, I have included them here because of their reliable, attractive evergreen foliage. This variety is pretty drought tolerant as long as a humus-rich habitat is provided at planting, with very large, rounded, glossy dark green leaves that are suffused purple in winter and mass up well. Magenta flowers appear on stout, red stems in spring.

BEST USES Effective as extensive low-maintenance ground cover on steep grassy banks or in partially shaded woodland; lovely planted with *Crocosmia* (montbretia)

DROUGHT TOLERANCE Good, once established
FLOWERS April to May
SCENTED No
ASPECT Any, in a sheltered or exposed position; full sun to partial shade
SOIL Any fertile, humus-rich, well-drained soil; add organic matter before planting
HARDINESS Fully hardy at temperatures down to -15°C/5°F; needs no winter protection
PROBLEMS Slugs and snails can damage young growth
CARE Remove brown or damaged foliage in spring; remove faded flower heads after flowering
PROPAGATION Division in spring or autumn; root cuttings of young rhizomes that have at least two leaf growths in autumn, after flowering is over

Cotoneaster dammeri ♀

⬆ 20cm/8in ↔ 2m/6ft   **EASY**

Originally from China, this vigorous, neat, spreading, prostrate evergreen shrub forms a low, dense matting of rounded shiny leaves, bearing simple pretty white flowers with pale yellow stamens that stud the long branching stems in early summer. Prolific small clusters of round red berries are produced in autumn.

BEST USES Works incredibly well as dense evergreen ground cover, useful for covering awkward banks or sloping verges; ideal for the cottage garden or wildlife garden, as birds love the berries

DROUGHT TOLERANCE Excellent, once established

FLOWERS June, but grown mainly for evergreen foliage and autumn berries

SCENTED No

ASPECT South, west or east facing, in a sheltered or exposed position; full sun to partial shade

SOIL Any fertile, well-drained soil; add organic matter before planting

HARDINESS Fully hardy at temperatures down to -15°C/5°F; needs no winter protection

PROBLEMS Woolly aphids; fireblight

CARE Prune back dead or damaged stems after flowering to restrict shape and spread

PROPAGATION Greenwood cuttings in early to late summer; layering in spring

Juniperus horizontalis
Creeping juniper

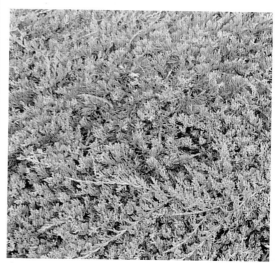

⬆ 30cm/12in ↔ Indefinite   **EASY**

This North American prostrate silver shrub is a very beautiful evergreen coniferous plant that has a horizontal, spreading habit. It has aromatic, silver-grey needled foliage, which adopts a purple hue in autumn, and small oval cone-like fruits. It is a very elegant, understated shrub, low on maintenance but high in visual appeal.

BEST USES Great for sloping banks or difficult-to-reach areas, where it will slowly spread, suppressing weeds and making an ornamental space of silvery grey shadow; also very salt-tolerant so worth trying in coastal gardens

DROUGHT TOLERANCE Excellent, once established

FLOWERS Grown for its foliage

SCENTED Aromatic leaves

ASPECT South, west or east facing, in a sheltered or exposed position; full sun to partial shade

SOIL Any fertile, well-drained soil, especially dry, chalky soil

HARDINESS Fully hardy at temperatures down to -15°C/5°F; needs no winter protection

PROBLEMS Aphids; honey fungus and twig blight

CARE Needs no pruning

PROPAGATION Take young shoots (still green at the base) and treat them the same as semi-ripe cuttings in late summer, autumn or winter (those taken in winter will need bottom heat)

Lonicera pileata
Privet/Box-leaved honeysuckle

⬆ 60cm/24in ⬌ 1.5m/5ft   EASY

This very useful, dense spreading evergreen shrub from China is fairly fast-growing and has small, glossy, lance-shaped leaves. The scented, small creamy-white flowers in spring are followed by attractive translucent purple berries. It's a great plant for offering interest all year round and is reportedly deerproof, so will make an attractive hedging solution if you are plagued by muntjacs munching your garden greenery.

BEST USES Makes very effective ground cover; good as a low formal hedge or a mounded, snaking hedge for contemporary gardens (but cut off the flowers to keep it dense as a hedge); ideal for coastal gardens as shelterbelt planting

DROUGHT TOLERANCE Excellent, once established
FLOWERS April to May, but grown mainly for foliage
SCENTED Lightly scented flowers
ASPECT Any, in a sheltered or exposed position; full sun to partial shade
SOIL Any fertile, moist, well-drained soil; add organic matter before planting
HARDINESS Fully hardy at temperatures down to -15°C/5°F; needs no winter protection
PROBLEMS Powdery mildew; can suffer from unsightly dieback
CARE Clip in spring or late summer or as often as needed to maintain size and shape as hedging
PROPAGATION Semi-ripe cuttings in late spring to summer (must be kept frost free); hardwood cuttings in autumn to mid-winter

Pachysandra terminalis
Japanese spurge

⬆ 20cm/8in ⬌ Indefinite   EASY

This low-growing evergreen perennial from Japan and China really earns its stripes for tackling those problem areas where very little else will grow, such as dry shade. It has very attractive, shiny, dark green toothed leaves and short stems bearing small spikes of brilliant white flowers in spring. It survives on a meagre diet (but can be invasive in very moist conditions), spreads fairly rapidly and offers year-round interest.

BEST USES Makes a very attractive, glossy evergreen carpet in a woodland setting or shady area in the garden; also well worth trying under trees and shrubs

DROUGHT TOLERANCE Good, once established
FLOWERS May to June, but grown mainly for foliage
SCENTED No
ASPECT South, west or east facing, in a sheltered or exposed position; full sun to partial shade
SOIL Any fertile, humus-rich, well-drained soil; add organic matter before planting
PROBLEMS None
CARE Cut back in spring or late summer to maintain desired size and spread
PROPAGATION Semi-ripe cuttings in summer or autumn; division in spring

Ruta graveolens 'Jackman's Blue'
Common rue

⬆ 60cm/24in ⬌ 60cm/24in   EASY

This delightful evergreen shrub forms a neat, rounded hummock. It has delicately cut, pungent leaves of the most elegant blue-green colour, almost metallic in their appearance, and in summer it bears insignificant pale yellow flowers. Forget the flowers – it is really grown for its attractive foliage: a divine architectural plant. It was traditionally used as a remedy for rheumatism, but the leaves can cause skin irritation, especially when handled in full sunlight.

BEST USES Very effective studding the front of a gravel or Mediterranean garden and looks equally attractive planted in multiples in low containers in a city garden or in the herb garden

DROUGHT TOLERANCE Excellent, once established
FLOWERS June to September, but grown mainly for foliage
SCENTED Aromatic leaves
ASPECT South, west or east facing, in a sheltered or exposed position; full sun to partial shade (the leaf colouring will be less pronounced in shade)
SOIL Any fertile, well-drained soil, including acid soils
HARDINESS Fully hardy at temperatures down to -15°C/5°F; needs no winter protection
PROBLEMS None, though can be prone to root rot
CARE Cut back by half in early spring to encourage new, bushy growth; trim lightly after flowering
PROPAGATION Sow seed in a cold frame in spring; greenwood or semi-ripe cuttings in summer or autumn

Uncinia rubra
Red hook sedge

⬆ 25cm/10in ⬌ 30cm/12in   EASY

This handsome, dwarf, evergreen perennial grass from New Zealand has shiny mahogany red foliage that looks good for most of the year, though the young foliage is more vivid than the mature leaves. It spreads by rhizomes, which probably accounts for its drought tolerance, and produces blackish spiked panicles in summer. The bright colouring in winter will add a splash of cheer to the gloomiest corners of the garden while also providing architectural shape, and partial shade will enhance its drought resistance.

BEST USES Useful in large swathes as ground cover; also very effective in containers, so ideal for a chic roof garden or shaded city courtyard

DROUGHT TOLERANCE Good, once established
FLOWERS July to August, but grown mainly for foliage
SCENTED No
ASPECT Any, in a sheltered or exposed position; full sun to partial shade
SOIL Any fertile, moist, well-drained soil; add organic matter before planting
HARDINESS Frost hardy at temperatures down to -5°C/23°F; needs winter protection
PROBLEMS None
CARE Cut back dead or damaged growth after flowering to keep the foliage display at its best; cut unruly clumps back hard to ground level
PROPAGATION Sow fresh husks at 13°C/55°F in spring after risk of frost has passed; division in spring or late summer

Coronilla valentina subsp. *glauca* 'Citrina' ♊

⬆ 80cm/32in ⬌ 80cm/32in   **EASY**

An indispensable, compact evergreen shrub that is hardier than you might imagine, though perhaps it is not so surprising when you learn it originates on the cliffs, woodland and scrubland of Europe and north Africa. It has very lovely tactile grey-green leaves with lightly scented pea-like pale lemon flowers, smelling faintly of daffodils. I can't think of anything else that will flower so persistently through all the iron-grey winter months – reason enough to grow it.

BEST USES A cheerful and stylish option for winter containers in a sheltered, sunny spot (saving us from those horribly predictable winter pansies!); will do well in a mixed shrub border or cottage garden; also excellent for coastal gardens

DROUGHT TOLERANCE Excellent, once established

FLOWERS November to March

SCENTED Lightly scented flowers

ASPECT South or west facing, in a sheltered position with protection from cold winds; full sun

SOIL Any fertile, well-drained soil

HARDINESS Frost hardy at temperatures down to -5°C/5°F; needs winter protection in colder areas

PROBLEMS None

CARE Cut out damaged or dead wood in early spring or autumn, when the plant is dormant; cut very leggy specimens back hard to the base

PROPAGATION Sow ripe seed in a cold frame immediately or in spring at a minimum of 10°C/50°F

Erysimum 'Bowles' Mauve' ♊
Wallflower

⬆ 75cm/30in ⬌ 60cm/24in   **EASY**

This is a corking, shrubby evergreen perennial, found in Europe and North America. It has attractive narrow, grey-green leaves and bears a fair profusion of very pretty deep mauve flowers from late winter to summer. It can be quite short-lived, but no matter: it's an easy plant from which to take cuttings, and the flowering is so prolonged you wonder how it keeps going! This is a real value-for-money plant, not only because it is evergreen but also for the lengthy flowering period, which is a real bonus.

BEST USES Perfect for the hot, sunny traditional border, but equally at home in the Mediterranean garden; excellent for the cottage or wildlife garden as it is irresistible to bees; tolerant of coastal conditions

DROUGHT TOLERANCE Good, once established

FLOWERS February to September

SCENTED No

ASPECT South, west or east facing, in a sheltered or exposed position; full sun

SOIL Any fertile, well-drained soil; can be short-lived in heavy clay

HARDINESS Fully hardy at temperatures down to -15°C/5°F; needs no winter protection

PROBLEMS Slugs and snails

CARE Trim lightly after flowering to prevent it becoming scrappy and leggy

PROPAGATION Heeled softwood cuttings in spring or summer

Mahonia aquifolium 'Apollo' ♎
Oregon grape

⬆ 90cm/3ft ⬌ 1.5m/5ft   EASY

Mahonias are tough, taking all weathers without flinching. This evergreen dwarf variety has a spreading, low-growing habit, with the trademark large, polished, deep green, stiff, spiny-edged architectural leaves, suffused with bronze that turn to claret hues in winter. It produces dense grape-like clusters of yellow flower spikes in late winter or early spring. Though not as highly scented as other varieties, there is a very slight spring-like fragrance. Small, round, grape-like black berries follow the flowers.

BEST USES Offers year-round interest to the mixed border; will appeal to the wildlife gardener as the berries are a favourite with winter scavenging birds; can be grown in full shade, so perfect for city courtyards, roof terraces or woodland

DROUGHT TOLERANCE Excellent, once established

FLOWERS November to March

SCENTED Slightly scented flowers

ASPECT Any, in a sheltered or exposed position; partial to full shade

SOIL Any fertile, well-drained soil

HARDINESS Fully hardy at temperatures down to -15°C/5°F; needs no winter protection

PROBLEMS Powdery mildew and rust

CARE Remove dead or damaged material after flowering; no regular pruning needed

PROPAGATION Sow ripe seed in autumn; division in spring or autumn; semi-ripe cuttings in summer or early autumn; hardwood cuttings in winter

Sarcococca hookeriana var. *digyna* 'Purple Stem' Sweet/Christmas box

⬆ 90cm/3ft ⬌ 90cm/3ft   EASY

A very lovely, rhizomatous, evergreen dwarf shrub from China with a suckering habit. It is rounded and compact, with narrow, dark green leaves and attractive purply black stems bearing a profusion of rosy flower buds that open to reveal tasselled, sweetly scented, white flowers flushed pink at the base in winter to early spring. Rounded black berries follow the flowers. The foliage is handsome all year round and the fragrance incredibly cheering in the cold grey of winter. As shrubs go, it must surely rate 10/10.

BEST USES Invaluable for mixed borders; does well in large containers, so excellent for city gardens or small patios; a low-maintenance solution for banks or sloping areas when planted in multiples; perfect for north-facing walls

DROUGHT TOLERANCE Excellent, once established

FLOWERS December to March

SCENTED Highly fragrant flowers

ASPECT Any, in a sheltered position; partial to full shade

SOIL Any fertile, well-drained soil

HARDINESS Fully hardy at temperatures down to -15°C/5°F; needs no winter protection

PROBLEMS None

CARE No regular pruning needed; remove dead or damaged material after flowering

PROPAGATION Sow ripe seed in autumn; semi-ripe cuttings in summer

Atriplex halimus
Tree purslane

⬆ 2m/6ft ⬌ 2.5m/8ft   EASY

A friend gave me a rooted cutting of this native semi-evergreen shrub and what an unexpected treasure it turned out to be. Vigorous, dense and spreading, it has attractive, diamond-shaped silver-grey leaves about 5cm/2in long, with rather ordinary, tiny white flowers in summer that aren't always reliable and, frankly, are so insignificant that you barely notice them. The foliage seems to whiten in very hot weather, the plant taking drought without flinching. For year-round foliage interest, it really is a lovely plant, and not as widely grown as it should be.

BEST USES Indispensable in a Mediterranean garden as background foliage; combines well with prairie-style planting; ideal in coastal areas as an isolated shrub, and would make an effective informal hedge or windbreak when planted in number

DROUGHT TOLERANCE Excellent, once established
FLOWERS July, but grown mainly for foliage
SCENTED No
ASPECT South or west facing, in a sheltered position with protection from cold winds; full sun
SOIL Any fertile, well-drained soil
HARDINESS Fully hardy at temperatures down to -15°C/5°F; needs no winter protection
PROBLEMS Box blight and red spider mite
CARE Cut back to restrict size and shape in spring, removing dead or damaged material
PROPAGATION Softwood cuttings in summer

Buxus sempervirens 🎖
Common box

⬆ 5m/16ft ⬌ 5m/16ft   EASY

This is a very slow-growing, very dense-leaved shrub native to northern Europe. It has small oval evergreen leaves and greeny-yellow flowers in spring, so insignificant that you will hardly notice them. It is fabulous: versatile and low maintenance. We use it over and over again in landscaping projects as it is so structural.

BEST USES Excellent as a formal hedging plant, to give 'bones' to the winter garden and in parterres and knot gardens; prized as an architectural shrub, and in containers as topiary

DROUGHT TOLERANCE Excellent, once established
FLOWERS April to May, but grown for foliage
SCENTED No
ASPECT Any, in a sheltered or exposed position; partial to full shade
SOIL Any fertile, humus-rich, well-drained soil; add organic matter before planting
HARDINESS Fully hardy at temperatures down to -15°C/5°F; needs no winter protection
PROBLEMS Red spider mite; box blight and leaf spot
CARE Mulch with organic matter in early spring; clip hedges in mid- to late summer; prune topiary after risk of frosts has passed and apply a slow-release fertiliser (fish, blood and bone or organic matter)
PROPAGATION Semi-ripe cuttings in summer to early autumn

Cordyline australis 'Torbay Dazzler' 🎖
Cabbage palm

⬆ 10m/32ft ⬌ 4m/12ft   **EASY**

An evergreen, architectural, palm-like shrub from New Zealand that has tall, slightly arching, sword-like tough green leaves with cream edging and central vertical pink striping down the centre. Mature trees produce upright panicles of tall (up to 90cm/3ft), densely clustered, tubular creamy white flowers that are very sweetly scented, followed by whitish berries. It gives all-year interest and structure to the garden.

BEST USES Excellent in containers, adding exotic architecture to city patios or roof terraces; great for coastal gardens, being salt tolerant

DROUGHT TOLERANCE Excellent, once established
FLOWERS July to August, but grown mainly for foliage
SCENTED Highly scented flowers
ASPECT South or west facing, in a sheltered position; full sun to partial shade
SOIL Any fertile, well-drained soil
HARDINESS Frost hardy at temperatures down to -5°C/23°F; needs winter protection in colder areas
PROBLEMS None
CARE Mulch with organic matter in early spring
PROPAGATION Sow seed at minimum 16°C/61°F in early spring; pot up rooted suckers in spring

Elaeagnus × *ebbingei* 'Limelight'

⬆ 3m/10ft ⬌ 3m/10ft   **EASY**

This elegant, dense, spreading evergreen shrub originates from Asia and has polished, rounded, silvery green leaves that develop a pale buttery yellow blotching in the centre when mature. It produces small, but highly fragrant, whitish flowers in autumn. It is a handsome shrub all year round, providing good seasonal interest.

BEST USES Excellent grown as a single specimen in a mixed border; also adapts well as hedging or a formal wall climber; tolerant of coastal conditions, so ideal as a windbreak for seaside gardens

DROUGHT TOLERANCE Excellent, once established
FLOWERS October to November, but grown mainly for foliage
SCENTED Highly scented flowers
ASPECT Any, in a sheltered or exposed position; full sun to partial shade
SOIL Any fertile, well-drained soil
HARDINESS Fully hardy at temperatures down to -15°C/5°F; needs no winter protection
PROBLEMS Coral spot
CARE Cut back lightly in late spring to restrict size and spread
PROPAGATION Sow ripe seed immediately; semi-ripe cuttings in late summer to autumn; hardwood cuttings in late autumn to late winter

Euonymus fortunei 'Silver Queen' ♍
Spindle 'Silver Queen'

⬆ 2.5m/8ft ⬌ 1.5m/5ft   **EASY**

An attractive, bushy, evergreen perennial, originally from Asia, that is easy to grow as a small shrub or as a neat wall climber (when it can reach 6m/20ft in height). Insignificant pale green flowers are produced in summer and these are followed by pale pink berries, but it is really grown for its reliable evergreen foliage. It delivers high visual appeal combined with low maintenance.

> **BEST USES** Ideal for courtyards, entrances and around front doors or growing up low walls

DROUGHT TOLERANCE Excellent, once established
FLOWERS Insignificant; grown for its foliage
SCENTED No
ASPECT Any, in a sheltered or exposed position; full sun to partial shade (but the leaf variegation is stronger in full sun)
SOIL Any fertile, well-drained soil
HARDINESS Fully hardy at temperatures down to -15°C/5°F; needs no winter protection
PROBLEMS None
CARE Minimal; trim in early spring before new growth appears, and remove any plain green foliage to stop the plant reverting (the leaves returning to their plain colour)
PROPAGATION Greenwood cuttings in late spring; softwood or semi-ripe cuttings in late spring to late summer; hardwood cuttings in autumn to late winter

Griselinia littoralis 'Variegata' ♍

⬆ 2.5m/8ft ⬌ 2.5m/8ft   **EASY**

These fairly upright, bushy, evergreen shrubs from New Zealand have dense, glossy, appley green leaves with marbled edging and insignificant greenish flowers in spring, followed by purple fruits in autumn (but only if plants of both sexes are grown). The foliage is a firm favourite with flower arrangers.

> **BEST USES** Excellent as a handsome low-maintenance hedge or as a windbreak for shelter planting in coastal areas, where risk of frost is minimal

DROUGHT TOLERANCE Excellent, once established
FLOWERS May, but grown for foliage
SCENTED No
ASPECT South or west facing, in a sheltered position; full sun
SOIL Any fertile, well-drained soil
HARDINESS Frost hardy at temperatures down to -5°C/23°F; needs winter protection in all but the mildest areas
PROBLEMS Leaf spot
CARE Cut back lightly in late spring to maintain size and shape
PROPAGATION Sow seed at 13–18°C/55–64°F in spring; semi-ripe cuttings in late summer to autumn; hardwood cuttings in late autumn to late winter

Hedera helix 'Glacier' ♈
Irish ivy

⬆ 2.5m/8ft ↔ 2m/6ft   EASY

Ivies are from diverse habitats, including
Britain, Europe and Asia. This strong-growing,
self-clinging, medium-sized evergreen climber
has handsome, glossy, leathery, triangular, lobed
leaves that are mid-silver grey-green, slightly
marbled, with splashes of cream at the edges. It
has umbels of limey green flowers in autumn, and
shiny, round black berries in autumn to winter.

BEST USES Marvellous for covering long stretches
of unsightly fencing, where precious little else will
grow; very effective as evergreen ground cover in
woodland settings; the berries encourage foraging
birds and wildlife into the garden

DROUGHT TOLERANCE Excellent, once established

FLOWERS October to November, but grown mainly
for foliage

SCENTED No

ASPECT Any, in a sheltered or exposed position; full sun
to full shade

SOIL Any fertile, humus-rich, well-drained soil; add
organic matter before planting

HARDINESS Fully hardy at temperatures down to
-15°C/5°F; needs no winter protection

PROBLEMS None

CARE Trim back to restrict size and spread at almost any
time of the year

PROPAGATION Softwood cuttings in spring; semi-ripe
or hardwood cuttings in late summer to late winter;
layering at any time

Jasminum nudiflorum ♈
Winter jasmine

⬆ 3m/10ft ↔ 3m/10ft   EASY

A mainstay of the winter garden, but originally
from China, this arching deciduous shrub can
also be trained very effectively as a wall climber.
Bright yellow primrose-like flowers appear before
the small, narrow oval leaves on long, arching,
naked dark green stems in early winter and
spring. This is a tough contender, for all its
delicacy, and will tolerate a wide range of soil
and weather conditions. Espied in mid-winter
snow or when the cheery flowers are jewelled
with rain, its enduring popularity is no surprise.

BEST USES Grow in the shrub border or train
against a warm wall in a courtyard or city garden;
works well as a flowering hedge or shaped in a
formal garden; tolerant of coastal conditions

DROUGHT TOLERANCE Good, once established

FLOWERS January to March

SCENTED Scented flowers

ASPECT South, west or east facing, in a sheltered
position; full sun to partial shade

SOIL Any fertile, well-drained soil; add organic matter
before planting

HARDINESS Fully hardy at temperatures down to
-15°C/5°F; needs no winter protection

PROBLEMS Aphids

CARE After flowering in spring, cut back to strong buds
on lower growth

PROPAGATION Semi-ripe cuttings in summer;
hardwood cuttings in winter; layering in spring

Lonicera fragrantissima
Shrubby honeysuckle

⬆ 2m/6ft ⬌ 2m/6ft   **EASY**

This scented winter-flowering shrub comes from China. It is a delightful deciduous, arching shrub that can also be trained as a wall climber. Semi-evergreen in milder areas, it has rounded, light green leaves and highly scented, tubular, creamy white flowers that mature to a soft, butter yellow, borne on the downward-sloping bare stems in late winter to early spring. Small, attractive red berries follow the flowers. It fully lives up to its Latin name '*fragrantissima*': no winter border should be without it.

> **BEST USES** Invaluable for its winter display in a mixed or shrub border and also complements emerging bulbs in spring; can be grown against a warm wall; will do well for coastal gardens

DROUGHT TOLERANCE Excellent, once established

FLOWERS November to March

SCENTED Scented flowers

ASPECT Any, in a sheltered or exposed position; partial to full shade

SOIL Any fertile, well-drained soil

HARDINESS Fully hardy at temperatures down to -15°C/5°F; needs no winter protection

PROBLEMS Aphids; powdery mildew

CARE Low maintenance; remove damaged or dead stems after flowering

PROPAGATION Softwood cuttings in late spring to early summer; layering in spring

Lonicera × *purpusii* 'Winter Beauty' 🎖

⬆ 2m/6ft ⬌ 2.5m/8ft   **EASY**

This is one of the loveliest honeysuckles, from China again, bringing fragrance and flower into the garden in mid-winter, yet is not widely planted, perhaps because people don't realise there is a shrubby variety available. It is a deciduous or semi-evergreen spreading shrub, with a lovely mounded habit and light green oval leaves, and can be trained as a wall shrub. From winter to early spring, it has profuse clusters of white, lemony-perfumed flowers that have distinctive yellow anthers on bare stems before the leaves unfurl. Berries are rarely produced.

> **BEST USES** Grow against a sheltered, sunny wall, or near an entrance or doorway to appreciate the perfume on a grey winter's day; ideal for a city patio and lovely in an informal or cottage garden

DROUGHT TOLERANCE Good, once established

FLOWERS December to March

SCENTED Lightly scented flowers

ASPECT Any, in a sheltered or exposed position; full sun to partial shade (but flowers more profusely in full sun)

SOIL Any fertile, well-drained soil

HARDINESS Fully hardy at temperatures down to -15°C/5°F; needs no winter protection

PROBLEMS None

CARE In late spring, cut back straggling stems and cut out dead or diseased wood at the base of the plant

PROPAGATION Semi-ripe cuttings in late spring to early summer; hardwood cuttings in late autumn to mid-winter

Mahonia × media 'Winter Sun' ♟

⬆ 5m/16ft ⬌ 4m/13ft   EASY

A much underrated group of architectural ever-green shrubs from America and Asia, mahonias provide year-round interest. This formal, upright variety offers all its name suggests. The dark green, shiny leaves are spiny-edged and the lily-of-the-valley-scented erect spikes of densely clustered sunshine yellow flowers appear in late winter to early spring, followed by attractive bunches of bluey purple-coloured berries.

BEST USES Worthy of inclusion in any mixed shrub border; can also form an effective animal-proof barrier and is ideal for deterring burglars!

DROUGHT TOLERANCE Excellent, once established

FLOWERS November to March

SCENTED Lightly scented flowers

ASPECT Any, in a sheltered or exposed position; partial to full shade

SOIL Any fertile, well-drained soil

HARDINESS Fully hardy at temperatures down to -15°C/5°F; needs no winter protection

PROBLEMS Powdery mildew and rust

CARE Trim lightly after flowering and cut out any dead or straggly stems

PROPAGATION Sow ripe seed in autumn; division in spring or autumn; semi-ripe cuttings in summer or early autumn; hardwood cuttings in winter

GREENFINGER TIP *If the plant gets leggy, reduce its size by a third in early spring and it will send out new bottom growth*

Phormium tenax ♟
New Zealand flax

⬆ 4m/13ft ⬌ 2m/6ft   EASY

An evergreen, clump-forming perennial from New Zealand, this needs plenty of space. With no central stems to speak of, it has tough, rigid, sword-like leaves of dark green with glaucous undersides. Stiff, upright stems of muted red, fairly fleeting flowers are produced in summer. *P.* 'Sundowner' ♟ (with bronzed green leaves and pale orange margins) and 'Dazzler' (red and pink striped leaves) are attractive varieties. All show their best colour planted in full sun.

BEST USES Ideal for coastal gardens; looks fabulous in a gravel garden, and adds architectural form to the border all year round

DROUGHT TOLERANCE Excellent, once established

FLOWERS July, but grown mainly for foliage

SCENTED No

ASPECT Any, in a sheltered or exposed position with shelter from cold winds; full sun to partial shade

SOIL Any fertile, well-drained soil

HARDINESS Fully hardy/borderline at temperatures down to -15°C/5°F; may need winter protection in colder areas

PROBLEMS Mealybug

CARE Cut out faded flower spikes and dead growth in spring

PROPAGATION Sow seed at 18°C/64°F in spring; division in spring

GREENFINGER TIP *The leaves are extremely sinewy: you will need a sharp pair of secateurs*

Pittosporum tenuifolium 'Silver Queen' 🎖

⬆ 4m/13ft ⟷ 2.5m/8ft   EASY

This handsome, elegant, evergreen shrub from New Zealand has densely growing, pale green foliage with creamy margins and clusters of small purple flowers, smelling of honey, in late spring and early summer. They make excellent hedging, and respond well to clipping and shaping.

BEST USES Ideal for a warm, sheltered wall in a Mediterranean garden; equally at home in a cottage garden or city patio; bees and butterflies enjoy the flowers, so good for wildlife gardens; will tolerate coastal conditions

DROUGHT TOLERANCE Excellent, once established
FLOWERS May, but grown mainly for foliage
SCENTED Scented flowers
ASPECT South, west or east facing, in a sheltered position with protection from cold winds; full sun to partial shade
SOIL Any fertile, well-drained soil
HARDINESS Frost hardy at temperatures down to -5°/23°F; needs winter protection in cold areas
PROBLEMS Leaf spot and powdery mildew outside; red spider mite and scale insect when grown indoors
CARE Low maintenance; remove dead or damaged material in late winter or early spring
PROPAGATION Sow ripe seed at 15°C/59°F in winter; layering in early spring; semi-ripe cuttings in autumn

Viburnum tinus 'Gwenllian' 🎖

⬆ 2.5m/8ft ⟷ 2.5m/8ft   EASY

Evergreen viburnums are excellent foliage shrubs and this Mediterranean variety is no exception. It is one of the more compact, bushy cultivars with elegant, highly polished oval leaves and clusters of flattened flower heads. They are dark pink in bud, opening to pinky white flushed flowers from winter to spring, followed by copious numbers of blackish berries in autumn which are often present at the same time as the flowers.

BEST USES Provides year-round interest to a formal or cottage garden; ideal for encouraging wildlife and pollinating insects into the garden; will suit coastal conditions as a hedge or windbreak

DROUGHT TOLERANCE Excellent, once established
FLOWERS December to March; also grown for foliage and berries
SCENTED No
ASPECT Any, in a sheltered position with protection from cold winds; full sun to full shade
SOIL Any fertile, well-drained soil
HARDINESS Fully hardy at temperatures down to -15°C/5°F; needs no winter protection
PROBLEMS Aphids and viburnum beetle; honey fungus and leaf spot
CARE In early spring, cut out dead or damaged material and trim to maintain size and shape
PROPAGATION Sow ripe seed in autumn; greenwood cuttings in late spring to early summer; semi-ripe cuttings in mid-summer to autumn; hardwood cuttings in winter

Combating drought

Choosing the right plant

Experienced gardeners live by the maxim 'The right plant in the right place': start by choosing plants that suit your conditions. Planting a sun-worshipping plant in a shady, boggy corner of the garden is just asking for trouble. It won't like it and will die or grow meanly. In the same way, a bog plant that enjoys sitting in water will not respond well to dry, stony soil.

Drought-tolerant plants have evolved to cope and survive under the hardships of their specific environments. These may include poor, thin soil that is low in nutrients, rocky hill terrain in a hot climate, wild grasslands, or exposed coastal areas. They have been imported by plant hunters or travelled by natural migration from overseas, but it is the conditions in their lands of origin that give a plant its individual characteristics and coping strategies. They may have long taproots or tuberous roots that help them search for or store water, and plants that are naturally drought resistant are often easy to recognise by their leaves. These may be:

- silver or glaucous
- small, thin or narrow, to reduce loss of moisture
- hairy, waxy or leathery
- fleshy, to act as water storage reserves (in cacti and succulents)

Almost without exception, naturally drought-tolerant plants enjoy:

- well-drained soil that is dry and stony
- lighter soil, such as sand and loam
- low fertility soil, not improved with organic matter

Droughtproofing the garden

You may look at some of the plants in this book, read that they appreciate moist conditions and question their drought resistance. Determining drought tolerance is not a precise process. You can help some plants flourish in hot, dry weather even though they are not naturally drought tolerant: you can give them coping strategies.

Plants with lush, green foliage, fibrous roots and, more often than not, larger leaf surfaces (i.e. most perennials) are definitely not naturally droughtproof, but some can come through just fine with a helping hand. Not all herbaceous perennials will survive drought when assisted by drought-combative measures. It is simply not in their nature. Trial and error gets you there in the end, and in this book I have chosen perennials that have been tried and tested during water shortages and prevailed.

The following strategies will go a long way in helping plants that are not naturally drought tolerant to cope with those dry, hot summers. They will also improve your garden soil no end, as well as widening your choice of plants.

- Enrich your garden soil with plenty of well-rotted organic matter, garden compost or leafmould when planting. This will help to retain moisture in the soil, making it available to the plants' roots when they need it.
- Apply an organic mulch about 5cm/2in thick at the base of the plants in spring every year to reinforce the measure above.
- Nurse plants in the first year or two with regular watering: this helps root development so they cope better with dry periods when established.
- Plant them in partial or full shade if they can tolerate this aspect (as woodland plants can): this reduces water loss from the leaves.

Drought-loving lavender, purple sage and artemisia

Improving the soil

Soil improvement is the first line of defence when it comes to droughtproofing the garden: it helps the soil to retain water. Experiment by pouring a watering can of water on to dry sandy soil. It will drain away quickly. Now mix that same sand with some organic matter and add water again. The water will be slow to drain away and the soil mixture will feel more spongy. You have just maximised its water-retention qualities. (See page 98 for more soil improvement methods.)

Mulching

The value of mulching cannot be overestimated. It simply means spreading a thick layer of organic material on the soil or around the base of particular plants and, in conjunction with soil improvement, it is one of the most effective ways to create a drought-tolerant empire. Mulching:

- traps moisture at plants' roots
- reduces water evaporation
- prevents weeds by blocking light to weed seed
- helps absorb warmth, resulting in better root establishment

ORGANIC MULCHES

- Home-made compost, leafmould and well-rotted farmyard manure add nutrients to the soil, improve its texture and act like sponges, holding water in the soil (see pages 98–9).
- Other organic mulches, such as bark or wood chippings and recycled cocoa shells, will help trap moisture and suppress weeds, and also look attractive, but do not provide nutrients or change structure. In the case of ornamental bark, the wood itself is not composted and actually takes nitrogen from the soil as it breaks down. You can overcome this by applying a nitrogenous fertiliser.
- Peat has no nutritional value. Its purpose is to aerate and hold moisture in the soil. Always wet peat before applying as it swells to absorb ten times its weight in water: adding water after application can leave your plants in danger of drowning! Its use is frowned upon by many gardeners as peat bogs are being destroyed in its supply.
- 'Dry' mulches are leaves, straw or bracken laid in heaps over the crowns of plants that need winter protection from cold and frosts.

INORGANIC MULCHES

- Black polythene suppresses weeds but is unattractive and impermeable.
- Permeable sheeting or mesh suppresses weeds and also allows air and water through to the soil.
- Floating mulches or fleece are placed over plants and act like a large cloche, protecting them in an insulated microclimate. (I once met a Canadian gardener who had clothed her borders in old sweatshirts. 'Well,' she said, when I asked her why, 'the book said to use fleece!')
- Pebbles and gravel are decorative. Applied over damp soil, they also help prevent water loss.

Mulching with some organic materials:

- increases the soil's ability to hold water
- improves the soil's nutrient levels, making plants stronger

As a bonus, mulching looks good. There is nothing more satisfying than seeing a rich, dark coating lying over your flowerbeds.

Mulching can be carried out at any time of year provided the ground is not frozen or bone dry (water well beforehand). Spring is the obvious time to mulch, before new growth is under way; once a year should be plenty.

Conserving water

Utilising all available water supplies resourcefully is essential. Fortunately, plants that originate in hot, dry climates actually prefer not to have too much water: planting these will immediately reduce the demand on your water resources. Silver-leaved plants, for example, have modified leaves, often small and narrow, that are covered in a fine cuticle or membrane (which incidentally gives them their characteristic leaf colouring) and this protects the leaves from losing as much water as their large, leafy herbaceous neighbours.

By far the most efficient and least expensive automatic watering method is a leaking hose system. Porous tubing, laid above or below ground in beds and borders, seeps water into the soil, where it is delivered directly to the plants' roots (and it can be timed to come on at night, when there is less wastage from evaporation).

Extra supplies of water can be collected in water butts, placed around the garden or next to downpipes to collect rainwater from gutters. Pure rainwater has very few dissolved salts, chemicals or fluoride and is excellent for watering plants. 'Grey' waste water, from baths, sinks and washing machines, can also be collected in butts fed from domestic downpipes and kept separate from the rainwater butts, but you may have to be careful here as this water will contain residues from cleaning products.

Aspect

Sun and shade

Does your garden face north, east, south or west? The direction your garden faces is referred to as its aspect. Gardens with a south-facing aspect will be in full sun all day (if the sun is out), while gardens that face north receive only indirect sunlight or will be in shade for much of the day, and are often damp spaces. Gardens that face east get light in the mornings, and west-facing gardens get all the afternoon and evening sun. City gardens, even if they are west or south facing, can have sunlight blocked by neighbouring buildings or tall trees, so may not receive as many hours of sun as a rural garden in an open position.

Stipa gigantea flourishes in full sun

All plants, without exception, need a certain amount of indirect or direct sunlight to survive. Some need full sun to grow at their best (as do many in this book), while others are happier in light shade: this is a basic consideration in siting any plant. It is important to be aware of your garden's aspect when choosing plants so you match them up with a suitable planting spot.

Most drought-tolerant plants of Mediterranean origin prefer full sun. It is what they are used to in their lands of origin (and a Mediterranean sun is more relentless than anything we have to offer here). There are also many drought-tolerant plants that can cope admirably with some shade (including hardy geraniums and vinca), but green, leafy woodland plants that are naturally shade loving (such as lamium and foxgloves) will need to be planted in light shade in humus-enriched soil if they are to come through dry spells.

Dry areas in the shade of large trees can be almost impossible to plant, but some of the ground cover plants, such as ajuga and vinca, will do well as long as plenty of organic matter is dug in or laid thickly around them from the first moment of planting and continued annually.

Shelter

Gardens with exposed aspects have little shelter from winds, frost, rain and snow (as well as being unprotected from direct overhead sunlight and warm, drying winds in summer). An area that is blasted by chilly north winds is no place for sun-loving plants. Many plants are vulnerable to the cold and wet of winter in their first two years, and providing shelter can help them settle down.

You may be lucky enough to have a brick-walled garden, or a garden space enclosed by fencing, hedges or mature trees and shrubs.

These boundaries will take the brunt of the weather. The base of a sunny wall will invariably have warmer soil at its footings, and rarely be exposed to any winds at all, providing a suitable environment for a less hardy plant that would not survive elsewhere in the garden. However, garden walls, sheltered though they may be, are also dry spots. If you are planting at the foot of a wall, choose a plant that likes dry conditions or incorporate plenty of organic matter before planting. In any garden, there are usually areas that have their own microclimates: the wise gardener makes good use of these weather sanctuaries and creates them where they are absent.

Seaside gardeners (see page 105) use shelter planting to increase the range of plants they can grow, planting hedges or installing man-made screens as a windbreak. Specifically sited to filter out the harsh, salt-laden winds that can be so destructive to plant establishment, the fence or hedge will divert winds from the lower growing perennials or climbers planted within its shelter.

Cities, because of the mass of heat-absorbing concrete, are normally warmer than exposed rural areas. In milder areas and city centres, frosts are unlikely to be a major vexation, but for those who live in cold regions and rural areas, hard frosts in winter can be a way of life. Greenhouses, polytunnels and conservatories offer a contrived environment in which to shelter plants that cannot withstand the harsher elements of our native climate, such as *Helichrysum petiolare* and callistemon. Cloches and cold frames will provide more limited protection and are especially good for sheltering fresh sown seed (see page 118). Dry mulching the crowns of recommended plants or tucking fleece, straw or dry sacking round plants are other ways of providing shelter from frosts and keeping the rootstock protected.

Soil

Plants need soil to grow. It anchors the plant in the ground, and provides the essential nutrients, air and water they need. Obvious though it may seem, there is more to growing a plant successfully than sticking it in any old ground and keeping your fingers crossed! First and foremost, use the right plant, right place principle: choose plants that will do well in the soil you already have (and give them the right aspect).

Types of soil

There are four common types of garden soil: clay, sandy, loam and chalky soils. They differ in their ability to retain water, their nutrient values and in their acidity or alkalinity. The most important thing about any soil is its texture (the composition of the soil's particles) because this governs the availability of air and water to your plants. Different plants prefer different soils and the type and quality of soil is a major factor in growing any plant successfully. Happily, many drought-tolerant plants will thrive in poor, hungry or sandy soils.

Sandy soil is light, dry and free-draining. It is coarse and runs through your fingers. Unfortunately, nutrients are easily washed away, so it usually needs to be enriched with organic matter, applied annually, to improve fertility and water retention. However, many Mediterranean plants prefer the poor growing conditions of sandy soil, so there is no need to improve the soil for them. Plants such as festuca, lavender, rosemary and verbascum that struggle in heavy, waterlogged soils will thrive in light sandy soil.

Echinacea purpurea, with its distinctive cone-shaped centres, thrives in well-drained soil

Nerines do well in poor, dry soil

Clay Roll a small lump of soil between your palms: if it turns into a sticky, damp ball, you have clay soil. It is wet and heavy to dig in winter but rock hard in times of drought, and plants may struggle to get their roots down into this dense footing. The plus side is that it is high in nutrients and supports a wide selection of plants, trees and shrubs. Ceratostigma, hardy geraniums and roses do well in clay as long as it isn't waterlogged.

Loam is the caviar of garden soils. It is a rich, dark brown colour, with a crumbly texture, high nutrient levels and good water retention. It is easy to dig and plant roots become easily established. Most of the plants mentioned in this book will do well in loam, including aster, anaphalis, briza, crambe and sternbergia, to name but a few.

Chalk (or limestone) soil can be a real challenge as there are many lime-hating plants, so planting choices are limited. Chalk soil is very free-draining and therefore lacking in fertility as nutrients get washed away. Adding organic matter annually is of huge benefit. Chalk can become as dry as a desert in summer and sticky and difficult to work on in winter. Hardy geraniums and knautia will do well in chalk soil.

A quick word about soil **acidity** or **alkalinity**, which can decide your choice of plants. This is known as pH and is measured on a scale of 1–14; soil with a pH of 7 is said to be neutral, below 7 is acid and above 7 is alkaline (you can buy a simple kit to test your soil's pH value). Most plants will grow on a pH between 6 and 7, so you don't need to worry about it too much, but some plants are acid-lovers and can't thrive in neutral soils; these are known as 'ericaceous' plants. Avoid them if you have neutral or alkaline soil.

Improving the soil

Plants take nitrogen from the soil as they grow. We disrupt this natural process by removing organic matter, such as prunings, lawn clippings and dead leaves which would normally break down to replace nitrogen in the soil, so it is essential it is replenished. A good indicator that soil is lacking in humus or organic matter is the absence of earthworms.

Although many of the naturally drought-tolerant plants in this book thrive in poor soil and will not benefit from being grown in soil that has been enriched, most plants (particularly native woodlanders) will appreciate added organic matter. Adding well-rotted manure, compost or leafmould improves the fertility, structure and aeration of the soil, which in turn aids good root growth and seed germination. It also prevents soil compaction – the formation of an impenetrable crust on the soil surface that impedes air and water and is the result of general cultivation or heavy treading. Improving the texture of the soil will also reduce the need for constant watering: organic matter acts like a sponge, helping the soil to retain water more efficiently.

Digging in organic matter isn't the only way to improve soil. Adding horticultural grit or small-sized gravel is invaluable for soils that are very heavy or compacted, such as clay. The grit opens up the soil, allowing water and air to penetrate more freely. It helps soil to warm up more quickly, increases the rate of bacterial activity and improves water drainage, which is, of course, vital to the healthy growth of so many of the drought-tolerant plants in this book.

Soil improvement will make the soil easier to work with and provide a better growing environment for your plants. As a general guide, the following can be done in late winter or early spring:

- sandy soil: dig in organic matter
- clay soil: dig in grit; dig in organic matter or spread on the surface
- compacted soil: dig in grit, and organic matter if needed; rotovate to break up the compacted surface layer
- wet soil: dig in grit or gravel; if the soil is very wet, consider digging soakaways, ditches or drains, or plant in raised beds

Organic soil improvers

Farmyard manure Make sure it is well rotted, or it will be too strong and damage your plants: it needs to be soot black and have very little smell. It is readily available in rural areas, or you can buy it by the bag at garden centres.

Apply annually, ideally in autumn on clay soil or late winter or early spring for sandy soils, before plants begin growth. If your soil is very heavy, compacted or waterlogged, dig it into the top 30cm/12in of soil in the first year to speed up the process. Otherwise, dig it in or just spread a thick layer of 5–7cm/2–3in around favoured shrubs, trees or perennials and let nature do the work.

Mushroom compost is a marvellous soil conditioner, adding nutrients such as nitrogen, phosphorous and potassium to the soil and trapping moisture. It is easy to handle and has the added bonus of slightly raising the soil pH.

Apply as a mulch, spreading a layer 5–7cm/2–3in thick on the soil (20kg/40lb will cover 1sq m/1sq yd) in autumn on clay soil or late winter or early spring for sandy soils. Leave it on the soil surface as it breaks down very quickly. A word of warning: mushroom compost contains concentrated amounts of soluble salts and nutrients and is not to be used for ericaceous plants.

Leafmould A woodland floor is covered with a rich, textured, crumb-like soil formed by years of decaying leaf fall. Rotted leaves are marvellous for adding bulk to the soil and providing a humus-rich soil full of micro-organisms and beneficial bacteria.

Collect fallen leaves and put into a heap, a wire mesh bin or black bin liners, keeping them separate from the rest of the compost as they take longer to rot down. Water the leaves if they are dry and leave them for a year or two.

Add a handful of leafmould to boost the bulk of your garden compost, use separately to mulch shrubs and herbaceous perennials or dig in as soil improver for sowing and planting. Very well-rotted leafmould (2–3 years old) can also be used for compost or seed sowing, mixed with equal parts of sharp sand, loam and garden compost.

Green waste soil improver is a coarse-textured material derived from domestic green waste, such as grass clippings, plants and prunings, which is composted by local authorities. It is designed to improve soil structure and is an environmentally friendly option, though not necessarily organic.

MAKING COMPOST

There are two composting methods. A 'hot' heap is made all in one go and needs a large quantity of composting material; it heats up and decays rapidly, and will be ready for use within a few months. 'Cool' heaps are built up in layers as material becomes available and will take a year or so to decay.

Hot heaps Make a base layer of woody plants or twigs to aid air circulation and drainage. Fill the bin or build the heap with layers of well-mixed green and brown material, watering as you go. Cover and leave to heat up. After a couple of weeks turn the heap, mixing well, and add water if it has dried out. Leave for some months until it reaches the desired texture. It should smell earthy, not

dank and rotten; sometimes compost can be lumpy and twiggy, but is still ready to spread on the garden even if it hasn't reached the brown crumbly stage.

Cool heaps Start with a base layer as above and add enough mixed composting material to make a 40cm/16in layer above this. Cover. Keep adding balanced ingredients until the bin is full or the pile is too big to be comfortably handled. Leave alone. After about a year, check to see whether the layers are fully rotted. If the top layer is still dry, but the bottom is ready, use the compost that is rotted and re-mix the drier, less decayed material back into the heap, add water, replace lid or cover and wait a few months more.

Garden compost is plant matter from the garden that has been left to decay. It is an efficient (and free) means of recycling garden waste. Any once-living plant material will compost; avoid substances that don't break down (ashes, tins) or attract vermin (food). Getting the right balance of compost ingredients improves with experience, but use roughly equal amounts of green and brown plant material.

Green material includes grass clippings, vegetable peelings, spent bedding plants, soft green and non-woody prunings as well as coffee grounds or teabags. These all rot quickly. Do not add perennial weeds such as dandelions, bindweed and buttercups. Nettles are the exception: being very nitrogenous, they act as compost accelerators.

Brown material includes cardboard, crushed eggshells, shredded paper, straw, wood shavings and tough hedge clippings. Woody plant material is slower to decay but bulks up the finished compost (chop it up to speed the process). Fallen leaves can be composted but are better used to make leafmould as they decay very slowly.

Compost bins or heaps (which can handle more material) can be hidden in any corner of the garden. Heaps are best sited directly on the soil or grass, in full sun or partial shade. All compost bins or heaps require lids or a cover of some sort – an old bit of carpet will do in most cases. Running several heaps or bins at once will ensure a constant supply of compost and you can use one while the other is still rotting down.

Damping down a compost heap

Growing

Planting

It is important to give your plants as much help as possible when planting, so that they establish strongly and healthily from the start. Make sure the ground is well prepared by incorporating plenty of organic matter into the soil. This will enable it to retain moisture more efficiently: even plants that are not considered drought resistant can survive in an improved soil, albeit they may be a bit the worse for wear at the end of long weeks of extended drought. All perennials, shrubs and trees benefit from this procedure except the silver-leaved plants, which thrive in poor soil.

When planting up individual plants, water the potted plant well before planting and add some well-rotted organic matter or compost to the bottom of the planting hole. I like to puddle each plant in with a good measure of water that fills the hole; let this drain away before backfilling the hole. This will ensure the roots of the plant get a thorough wetting. Bare-rooted plants can be soaked for half an hour in a bucket of water before planting.

Watering

Newly planted perennials, trees or shrubs need regular watering to help them get their roots down and establish, but it is better to water once generously than more often meagrely: frequent watering can encourage plants to root shallowly near the soil surface, making them dependent on being watered and vulnerable in times of drought, while measured watering encourages strong, deep roots, so plants are able to withstand water shortage more easily. This approach will equip plants to get through the worst of a dry, hot summer.

Silver-leaved plants like to be watered at the base as splashing their leaves can make them vulnerable to disease. Don't waste water: direct water to the roots of plants, where it can be used most efficiently, and water early in the morning or in the evening, when less water will be lost through evaporation.

Feeding

Plants take nutrients and water out of the soil, and what is taken out must be replaced. Adding organic matter to the soil will maintain its fertility as well as improving soil structure and water retention, so mulch annually in spring (see page 94).

Fish, blood and bone, pelleted chicken manure and seaweed all feed the plant rather than improve the structure of the soil and are quite concentrated. Read the instructions carefully as you can harm a plant by overfeeding.

Comfrey liquid smells awful, but is cheap and very effective. Fill a drum or water butt with comfrey leaves and cover with water and a tight-fitting lid. Alternatively, put the leaves in a barrel without water and weigh down with a brick; this does away with the smell. Leave for 4–5 weeks in warm weather, 2–3 months in winter. Drain off the rich, dark liquid and dilute with water at a ratio of 15/20:1. Use as often as required. Rotted and fresh comfrey leaves are both excellent accelerators of the decaying process in the compost heap.

Granular fertilisers are comparatively fast-acting and are useful for a shot of nitrogen, which is very good for vegetative or leafy growth, or potassium, which encourages good flower or fruit formation. These fertilisers are a quick fix and do not improve the texture, structure or water-retention of the soil.

Plant husbandry

Please take time to look at your plants closely. Notice those that are growing away well, with plenty of healthy leaf growth or flower: you have planted them in exactly the right spot and looked after them well. Those that are struggling or looking feeble a couple of weeks after planting are simply not happy.

Ask yourself why. Too little sun? Or perhaps the leaves are scorched, indicating that too much sun or cold winds are affecting the plant. Are the leaves wilting and brittle? This is a sure sign they could do with more water. Perhaps the leaves are drooping? You may have planted them in too damp a soil and they need better drainage. Plants are very vocal: if you look carefully they will often tell you what they are lacking by their overall physical demeanour. With experience you will learn to hear what they are saying.

Planting with perennials

Perennials are a hugely varied group of plants, living at least two years and often much longer. Strictly speaking they include woody perennials such as evergreen or deciduous trees and shrubs, herbaceous perennials, and tender perennials that require some winter protection. The one thing they have in common is they all bring flowers and foliage into the garden.

Herbaceous perennials (just called perennials in the trade) are usually soft stemmed (as opposed to woody) and the top-growth dies back in winter while the roots live on under the ground. They can have bulbous, rhizomatous, fibrous or tap roots. Roots are often a clue to drought tolerance. Taprooted plants delve deep in search of water, whereas plants with spreading, fibrous roots absorb water nearer the surface and are more prone to drying out. Some perennials have a carpeting nature while others are tall and showy. They can be mixed with ornamental grasses, shrubs and trees in the flower border, but can also be grown in containers or wildlife gardens as the mood takes you.

Climate change has brought drier weather for longer periods and many traditional border plants simply cannot cope with extended periods of drought. With diminishing rainfall, gardeners need to reconsider their approach to gardening, because the lush flower borders of yesterday will prove increasingly difficult to maintain. Mixing drought-tolerant stalwarts with some old favourites can be an innovative, exciting way to inject new life into our borders or containers, helping us move with the times both climatically and creatively.

Caring for perennials

Perennials flower at different times of the year, but most die down with the onset of autumn or winter weather. They are generally cut down to ground level in late winter or early spring or die back naturally. You will have to judge which ones to cut back and when. No great science here: remove flowering stems if they look awful, trim back dying leaves to keep plants looking tidy, and deadhead prolific self-seeders if you wish to inhibit their spread.

Hardy geraniums can be deadheaded to induce a second flowering in the same year, rather like many roses, while alliums grace the flower border with the skeletal beauty of their seed heads long after the flowering is over. Some plants disappear underground in winter and you might be wise to mark their position so you don't dig them up by mistake.

Perennials need regular watering to help them establish but then need little attention or feeding, apart from an organic mulch in early spring to keep growth vibrant and healthy.

Perennial grasses

Perennial grasses are a diverse group of plants offering a range of foliage colour, textures and forms and are suited to a wide range of growing conditions. They add unexpected elements of movement and sound: a soft breeze may make them ripple in soft undulating waves or send their seed heads rattling gently. Many of the ornamental grasses hold their shape beautifully through the winter months. Rather than cutting them back after flowering, let them stand; they add structure to a winter landscape and look quite magical touched by frosts.

Caring for grasses

Once established, most grasses are trouble free. Preparing the soil with organic matter before planting will help them establish more quickly, but they need little further attention apart from weeding.

Trim back perennial deciduous grasses to about 10cm/4in in February. You may cut them back in the autumn if you wish, but I find they overwinter better if cutting is left until late winter; this also prompts a good growth spurt in spring.

Evergreen grasses don't take kindly to drastic cutting. Just remove any dead, bent or tatty leaves and prune out old flowering spikes in spring.

Planting with shrubs

Shrubs are woody-stemmed plants ranging in height from a mere 30cm/12in up to 3m/10ft or more (some shrubs, such as pittosporum, can be mistaken for small trees because of their eventual size). Shrubs can be evergreen or deciduous, often have gorgeous foliage and may have showy or fragrant flowers, sometimes followed by interesting fruits or berries.

Drought-tolerant grasses add texture, form and colour to any flower border

They are a vital addition to any garden, as they are low maintenance, offer year-round interest and establish good structure. When used as hedging they provide privacy or sanctuary. No garden should be without a complement of choice shrubs, planted with other shrubs or mixed with perennials in a border.

Caring for shrubs

Shrubs are generally low maintenance. All shrubs benefit from an annual mulch of organic matter in spring, to keep them healthy and to help retain moisture at the roots in dry weather, but they need very little watering once established. Some shrubs (such as French lavender) may be borderline hardy and need the shelter of a warm wall or the protection of fleece in winter but, by and large, hardy shrubs are reliable and most need no cosseting.

Remove any dead or damaged growth as it occurs. Shrubs that are grown as an evergreen hedge will need clipping at least two or three times a year. However, a shrub grown as a solitary focal point only needs occasional pruning to maintain its size and spread. Some (such as euonymus or some cotoneaster) can be trained as wall climbers, in which case they need pruning once or twice a year to ensure they grow in an upwardly mobile direction.

Container gardening

Pots of all description excite gardeners' imaginations. Terracotta pots can add to the Mediterranean atmosphere of a garden, but are equally appropriate for the cottage garden. A large urn makes an effective focal point in a flowerbed or border, and herbs are handy in pots outside the kitchen door. In a small patio or courtyard, growing plants in pots is a practical solution to shortage of space. Small olive trees, climbers, grasses, lavender, rosemary and roses can all be successfully cultivated in containers.

When planting in containers:

- choose frostproof pots
- lay crocks at the bottom of the pots over the drainage holes before filling with compost
- use a suitable compost
- mix swelling water granules with the compost: these reduce the need for constant watering
- cover the soil surface with a mulch of leafmould, mushroom compost, decorative gravel or bark to help prevent water evaporation

Plants in pots dry out much more quickly than those planted in the ground, as containers absorb and lose heat quickly. The compost should feel damp to the touch, but not waterlogged. Water generously when necessary (this could be every other day but, if you have used water granules, once a week should be sufficient) and ensure the water is soaking through to the roots. Generous watering aids good deep root development, making your plants better able to cope with dry weather. Sometimes the compost gets a dry cap or crust, preventing water reaching the roots: a drop of washing-up liquid in the watering can will break the surface tension.

The soil in pots and containers needs topping up annually. Plants soon use up the nutrients in their limited portion of soil, so you will need to provide a nutritious diet. Use pellets of slow-release fertiliser, mulch annually with mushroom compost or apply a liquid feed. Once a year is all that is needed, but it is too often forgotten.

Container plants are susceptible to vine weevil, so check your pots regularly. A drench (available from garden centres) that will treat and prevent vine weevil infestations can be applied to pots.

Sempervivums in a terracotta pot add interest to a patio

Coastal-friendly: a cardoon (*Cynara cardunculus*)

If your plant gets too big for its pot, you may find roots straddling the sides or growing through the drainage holes: consider repotting into a larger pot. However, large climbers or shrubs grown in containers will never reach the full height or spread given in the plant description. Roots are restricted in a pot and the top-growth is limited by the reduced root spread below.

Coastal gardening

Gardeners on or near the coast or seaside have to deal with some very difficult physical challenges. Strong salt-laden winds are a particular problem and most coastal gardeners will understand the need for shelter planting (see page 96). Plants that survive with ease in other areas will struggle or die when faced with such unrelenting coastal conditions, but droughtproof plants are an absolute godsend to seaside gardeners.

There is far less risk of frost in coastal areas, so some of the more tender plants can be grown, and the cloud cover can be more sparse than inland, giving more hours of direct light. Some areas, such as Cornwall and parts of the Irish coast, are influenced by the Gulf Stream, resulting in a local milder climate. Here, gardeners can grow many species of less hardy plants that would not survive in colder neighbouring regions. Many of these are naturally drought tolerant.

There are numerous perennials and shrubs that can cope with coastal conditions commendably well (see page 124). Achillea, agapanthus, armeria and bergenia are just a few of the natural drought busters that positively embrace seaside locations. They may still need protection from the worst of the wind in the first year or two, but thereafter they should prove reliable.

Hardiness

Plant hardiness ratings are judged by different world temperature zones, but in the UK we generally use the RHS hardiness zones, which are explained here. Low temperatures can prevent certain plants from developing well; flower buds may fail to open, and leaves may fall prematurely. If the temperature falls below the plant's level of tolerance for any length of time, it will need some form of protection to prevent lasting damage. You may be surprised to learn that many drought-tolerant plants are just as hardy as many of our native perennials, despite originating in warmer climates.

Fully hardy: hardy to -15°C/5°F

This term describes climbers, perennials, shrubs, trees and fruit and vegetables that are naturally tough enough to withstand a lowest winter temperature of -15°C/5°F without suffering lasting harm. No extra mollycoddling from your good self is required. Plants that fall into this category originate in cold climates and are naturally adept at coping with cold, winds and frosts.

As a novice gardener I tended to grow plants that were considered fully hardy, having neither the time nor space to nurture more tender species. (That said, if you have the luxury of space that can act as winter shelter, whether a porch, polytunnel, conservatory or greenhouse, don't be put off growing some of the less hardy species.)

Frost hardy: hardy to -5°C/23°F

Plants in this category can withstand temperatures as low as -5°C/23°F. Once temperatures dip below this, especially if they stay low for any length of time, the plant may suffer lasting harm or even death. These plants need their roots protected from being frozen, and their top-growth from being terminally damaged by frosts. This is why you are advised to protect them with horticultural fleece (available from all garden centres) or provide shelter, such as a frost-free greenhouse.

Fleece can be packed around the base of a plant, to protect the roots, or wrapped around the plant itself and tied with garden twine, to provide protection for the leaves and stems. Mulch the crowns of vulnerable plants with dry leaves, fern fronds or straw, and wrap containers in bubblewrap or old sacking to give that extra layer of protection. The idea is to provide a winter layer, rather like a scarf, to take the brunt of the weather, so it can't affect the plant directly.

Borderline

Sometimes plants are referred to as being 'borderline' between two classifications, such as fully hardy and frost hardy. This means that they are teetering on the edge of a category and may be adversely affected by cold snaps. Even plants that are tough enough to have been labelled fully hardy can sometimes be affected by prolonged periods of bitterly cold weather in cold regions or exposed sites. You'll have to judge this borderline caper by your own knowledge of local temperatures and personal experience. If in doubt, provide cover.

However, more and more, I find that plants that are deemed frost hardy will survive a severe winter if I mulch the roots well in late autumn and site them against a warm, sheltered wall from the outset, protecting them from the worst of cold winds, rain and frosts. This has sometimes meant an emergency marathon around the garden, scraping snow from susceptible plants or providing fleecing for those that had looked like they might survive but, under the onslaught of cruel weather, are beginning to look decidedly feeble.

Occasionally I might lose one, but by taking a cutting earlier in the year or dividing the plant in spring, I have an insurance policy in case of a fatality. Eventually, experience will lead you to make the right decisions for your garden plants and you'll know when to take a risk – or not, as the case may be.

Half hardy: hardy to 0°C/32°F

Plants in this category will withstand temperature drops down to 0°C/32°F. They need protection with horticultural fleece and dry mulches, cloches and cold frames, or frost-free shelter, such as a porch, greenhouse or conservatory, where the temperature will never dip below the stated tolerance. This ensures that the plants' roots are protected from being frozen, or their top-growth being fatally damaged by snow or frosts.

Cloches and cold frames provide a microclimate for the plants and seeds sheltering under them, giving protection from ice, wind, snow and excessive wet. In spring, when the weather warms up, they also act as a mini-greenhouse, allowing plants to come on faster than those planted in open ground. Be aware that the use of these has to be carefully judged: they have no frost thermostat as in a greenhouse and tender plants can be damaged.

Frost tender: not hardy below 5°C/41°F

Frost-tender plants are those that are grown as annuals or that will not survive the severe winters of the British Isles without protection. Although you can protect a frost-tender plant from frosts, you will never make it hardy. The frost-tender rating will vary from plant to plant and is normally specified, but all frost-tender plants that are not grown as annuals will have to be overwintered in fleece, cloches or cold frames, or kept in a warm greenhouse or conservatory and moved outside in the summer months, once all risks of frost have gone.

Many of these plants will need a period of 'hardening off' or gentle acclimatisation to colder outdoor conditions before they can be moved outside (see page 118).

Drought-tolerant perennials, including erigeron, euphorbias and poppies, make effective ground cover in a Mediterranean-style garden

Problems

All gardeners have to deal with pests and diseases and you will very quickly learn the remedial action to take. There is not a garden plant in existence that won't fall prey to pests or diseases at some time, but good plant husbandry can prevent many ailments occurring in the first place.

TIPS FOR PROBLEM-FREE PLANTS

- Always buy healthy, vigorous plants from a reputable source as they will naturally have stronger resistance to disease.
- Look after your soil. Healthy growth depends on a healthy soil: the better the structure of the soil, the easier it is for plants to thrive (and the better they will combat periods of drought). Feed the soil rather than the plant: this leads to stronger growth and better resistance to pests and diseases.
- Encourage the natural enemies of garden pests. Naturally occurring predators such as birds, frogs, toads, shrews, hedgehogs, ladybirds and predatory ground beetles will all reduce the amount of pest infestation and help to maintain a healthy garden environment.
- Adopt organic garden practices. These are better for your garden and the planet, as well as being cheaper than chemical alternatives.
- Last, but by no means least, be vigilant. There is no substitute for this. If you keep a close eye on your plants, you will inevitably see problems before they race out of control. Aphids can soon smother plants but, if caught early, a mass infestation can be prevented. Similarly with leaf diseases, an early diagnosis can save more serious problems later.

Leaf and stem pests

It is often easier to identify pests than plant diseases. Sometimes there are obvious signs, such as unsightly chewed leaves, or you might catch a snail or slug red-handed. But often the culprit is nowhere in sight, so you have to put on your deerstalker and look carefully at the evidence to determine the problem. Here are some of the most common garden leaf pests you may encounter, with suggestions for dealing with them.

Aphids include pests such as blackfly, greenfly, lupin aphid (grey in colour) and whitefly. They are hard to eradicate completely because they occur in large numbers, never in isolation. They are all sapsuckers, feeding on the sap of young shoots, which causes the leaves and stems of a plant to curl and distort, and damages new emerging growth. This weakens the plant considerably. Aphids also excrete a sticky sugary substance, which is fed on by ants, so a column of ants advancing up the stems of your plants is a clue that you have got an aphid infestation.

The symptoms and remedies for blackfly, greenfly, greyfly and whitefly infestations are much the same. The pests are about 3mm/⅛in long, with transparent greenish, black or whitish bodies. They normally form a seething mass around the young stems of affected plants, where the growth is sappy and easier to attack. Greenfly occur almost everywhere outdoors and affect just about every plant you can think of. Whitefly are more common indoors, in greenhouses and conservatories, and blackfly tend to favour broad beans and elder, so should not concern us too much here. Greyfly are found on lupins, and can ruin a plant in a matter of days. **Solution** Spray as often as needed with a few drops of washing-up liquid mixed with water. For the voracious lupin aphid, remove infected growth immediately, before spraying in the evenings only. Alternatively, use a proprietary pest spray.

Capsid bugs are pale green sap-sucking bugs some 6mm/¼in long that cause distortion to leaf tips and pepper leaves with tiny holes. Flowers may be damaged or distorted.

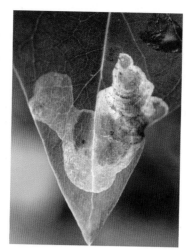

LEFT TO RIGHT Ladybirds are natural pest controllers; leaf miners inflict unsightly damage on foliage; the white foam of cuckoo spit

Solution Destroy adults by hand and winter wash trees with a proprietary product to destroy eggs. As a preventive measure, keep the area debris free. Alternatively, use a proprietary insecticide.

Caterpillars eat the leaves of many plants, leaving unsightly holes and notching. Silky webbing on leaves is a fair indicator they are present, but they are not shy and most are easily visible to the naked eye.
Solution All species are dealt with in the same way. Pick off by hand or remove the leaves and destroy them. If the infestation is large, use a proprietary insecticide.

Earwigs are brown insects, about 2cm/¾in long, with a set of pincers at the rear. They tend to feed at night, nibbling at leaves and flowers, so it is hard to prevent them inflicting damage unless you are prepared to stake out the garden at midnight.
Solution The damage is marginally unsightly, but not worth getting too concerned about. Leaf growth generated in the growing season will disguise the odd chewed leaf.

Froghoppers are responsible for the common white spit-like froth (often called cuckoo spit) seen on the stems and leaves of garden plants, particularly lavender, in early summer. A small, green sap-sucking nymph is camouflaged inside the foam and can stunt and distort plant growth.
Solution The damage is minimal but you can wash them off with soft soap solution.

Leaf miners are the larvae of beetles, moths and flies. An infestation can be easily identified from the tell-tale brown and white tunnel patterns on the leaf surface.
Solution Pick off and dispose of affected leaves. If you have a garden bonfire, burn them. Alternatively, use a proprietary spray.

Mealybug are found on houseplants or anything grown in a pot in a greenhouse or conservatory. They are tiny pinkish grey sapsuckers, normally covered in a woolly coating. This is actually a waxy layer, which acts as protective armour and makes it difficult to eradicate them.
Solution Spray water mixed with a couple of drops of washing-up liquid on the plant to dissolve the wax coating (use as often as you like as it does the plant no harm). Or, use a proprietary insecticide.

Narcissus bulb fly is a hovering fly that looks much like a bee, only it comes in tan! It lays its eggs at the base of plants as the foliage dies back and the creamy coloured larvae burrow into the centre of the bulbs, killing the developing flower buds, so bulbs come up straggly and blind (no flower).
Solution Bulb companies often treat bulbs to prevent this and it is an unusual occurrence. If practical, check the base of plants for the eggs in spring. Dig up and dispose of affected bulbs and plant new bulbs somewhere else.

Narcissus eelworm are microscopic nematodes which inhabit bulbs of daffodils, bluebells and snowdrops. Initial symptoms are distorted and yellowed growth above ground; bulbs will eventually rot. If you suspect your plants are affected, cut through a bulb and look for concentric brown ringing. Eelworm spread through the soil to neighbouring plants.
Solution Dig up and burn or dispose of all affected plants. Do not replant the area with bulbs susceptible to eelworm for two years. Always buy healthy bulbs from a reputable source.

Onion flies are small grey flies that lay their eggs in spring and early summer at the base of vulnerable plants such as alliums. The worst damage occurs in the larval phase when the white maggots, which are about 1.25cm/½in long, tunnel into plant tissue.
Solution Lift and destroy affected plants, ensuring that the soil around the plant is also maggot free. Chemical controls to treat the soil are available from garden centres.

Phlox eelworm are microscopic eelworms inhabiting the stems and leaves of phlox. Symptoms include stunted, distorted stems, shoots dying back and failure to flower.

Solution Dig up and burn or otherwise dispose of affected plants. There are no chemical or biological solutions available.

Red spider mite take many forms, but the most common is a barely visible sap-feeding mite which is found on houseplants or pot-grown greenhouse or conservatory plants and is hard to spot. If you notice mottled leaves or sudden yellowing, with silky webs around the young leaves and tips of your plants, you probably have red spider mite.
Solution Spray the underside of the leaves with an up-turned hose from the start of the growing period to prevent colonisation. In a greenhouse, keep humidity levels high by hosing down inside it daily.

Phytoseiulus persimilis is a predatory mite that reproduces faster than red spider mite and preys aggressively on its eggs, young and adults. This is effective if introduced into the greenhouse or conservatory early in the growing season, at a minimum constant temperature of 16°C/61°F.

Red spider mite are very resistant to chemical sprays: almost daily treatment is needed for any real effect.

Scale insects' tiny shells or scales are found on stems or, more commonly, the undersides of plant leaves. Infestation may result in very poor growth. Keep a watchful eye, as large infestations can be hard to treat.
Solution A small attack is of negligible consequence, but a heavy infestation can be very damaging to a plant. Apply methylated spirit to the leaves, using cotton wool or a cotton bud, or use a proprietary organic or chemical spray.

Slugs and snails are the two most common causes of plant damage. Their appetite for leaf and flower material is legendary and remedies for dealing with them are

Vine weevils have a voracious appetite

infamously numerous (and vary greatly in effectiveness). Look for slime trails and leaf damage: leaves that are munched close to the bottom of the plant are likely to have been attacked by slugs; top leaf damage is more commonly inflicted by snails, who don't mind the climb!

Solution Fill a small plastic tray with beer or sugar water and place shallowly in the soil at the foot of the plant. Slugs and snails are lured to this 'slug pub' as effectively as the local drunk at Happy Hour and drown in the sugary liquid.

Biological controls are available. Alternatively, sprinkle slug pellets sparingly around the base of your plants. These will kill the slugs or snails on contact.

Viburnum beetle larvae are creamy white and about 8mm/⅓in long. In early summer they eat viburnum leaves, just leaving a tracery of veins, before dropping to the ground to mature. The adult beetles start grazing on the leaves in July and August, laying their eggs on young shoot tips to hatch out the following spring.

Solution The damage to the leaves is unsightly, but rarely serious. Remove all affected leaves and dispose of them, thereby getting rid of the eggs as well. Place bird feeders near the plant to encourage birds to feed on the larvae.

Vine weevil most commonly affect plants grown in containers and pots, although they can also damage plants growing in open ground. The adult beetle is easily recognised by its long pointed snout; it feeds voraciously on the foliage of herbaceous plants and shrubs, causing unsightly notching along the leaf edges. This is unattractive, but rarely proves fatal. The real damage is done by their etiolated offspring: the larvae are hatched at the roots of the plants the parents are feeding on and the infants' diet is the plant's roots.

The first noticeable sign of the larvae may be yellowing of the leaves, poor growth and a wilting plant that does not respond to watering. Often there is very little warning; it is not uncommon to see an otherwise healthy plant suddenly keel over before you realise you have an infestation.

Solution A microscopic parasitic eelworm is available that enters the larvae and releases bacteria that kill them. To add to its effectiveness, it keeps reproducing inside the dead grub. Alternatively, a chemical drench (available from garden centres) can be watered over pots.

Woolly aphids affect pyracantha and cotoneaster. The most common symptom is white woolly or fluffy patches that house the young aphids, which are pinky brown. They are normally found in old pruning wounds or in cracks in the bark.

Solution If the infestation is minor, paint with methylated spirit and scrape them off or scrub with a hard brush and soapy water. Alternatively, spray with an insecticide in spring.

Plant diseases

Plant diseases can be a lot harder to spot than some of the common garden pests: it is not always easy to recognise that a plant is sickly, let alone find the cause of the trouble. Following some simple guidelines can often prevent nasty conditions spiralling out of control, so here are a few tips for keeping your plants disease free.

TIPS FOR DISEASE-FREE PLANTS

- Grow disease-resistant varieties wherever possible.
- Plant the right plant in the right place: a shade-loving plant will struggle in a sunny position and be more vulnerable to infection.
- Keep plants well watered and mulched to prevent roots drying out: caring for plants with regular watering and feeding equips them with stronger disease resistance.
- Improve the airflow round plants and avoid planting too closely together.
- Practise good hygiene in the greenhouse and garden: this is an obvious deterrent to plant sickness. Clear up fallen leaves, where fungal diseases can overwinter, and make sure pots are washed before potting up new plants or seedlings.

However, with the best will in the world, you are likely to meet one or two unpleasant diseases, no matter how good your garden hygiene, so here are some of the more common plant diseases and their solutions.

Blackspot is a fungal disease which affects roses and is most commonly seen during prolonged bouts of wet weather, as it is typically spread by water splashes and rain. Black markings appear on leaves, which then drop off prematurely. New roses have been bred with an in-built resistance to blackspot, and it is worth pursuing these. **Solution** Remove all infected foliage and dispose of it. Spray skimmed milk mixed with water at two-weekly intervals to keep the problem in check. Alternatively, use a proprietary spray.

Botrytis (grey mould) is a very common fungal disease that strikes weak plants and flourishes in damp and poorly ventilated conditions. It can survive on live plants and on dead or decaying plant tissue, and affects bulbs, annuals, perennials, trees and shrubs. Grey, fluffy mould containing spores is evident and will disperse in the air, spreading the spores, if handled carelessly. It can also enter plants though wounds, causing leaves to brown and soften before becoming covered in the grey mould.
Solution It is difficult to control as it is spread through the air, but good growing conditions and plant hygiene do much to prevent it. Remove all affected parts, cutting back to healthy growth, and burn or dispose of infected material, but do not add to the compost heap. Alternatively, use a fungicidal spray.

Box blight is a fungal disease which affects all box species and occurs especially in damp and poorly ventilated conditions. Symptoms include pink spores on the underside of leaves and grey fungal patches on leaf surfaces. Blackened and discoloured leaves fall from the plant as the disease spreads to the stems.
Solution Improve hygiene by removing all decaying debris or dead leaf material from the base of the plants and in the close vicinity. Prune out infected areas to prevent further spread. It is impossible to improve ventilation in box planted closely as hedging. Water plants from the base rather than above. Use clean clipping or pruning tools. There are no chemical or biological solutions available to home gardeners.

Coral spot is a fungal disease that affects trees and shrubs, and lives on both living and dead material. The spores enter the plant through wounds from cutting, pruning or weather damage; by the time the pinky orange lumps are spotted, woody stems are dead. It is a mild condition, which normally only attacks a stressed or ailing plant, but drought conditions are just the sort that can trigger it.
Solution Prune out affected shoots and stems to healthy tissue and burn or dispose of infected material – but do not compost.

Downy mildew is a fungal disease that is less common than powdery mildew, causing yellow to brown blotching on the upper surface of the leaves, while a white fuzz may develop on the undersides. Left unattended, a whole plant may discolour and die.
Solution Remove and destroy all infected leaves and avoid overhead watering which can exacerbate the problem. Fungicidal sprays may be used, but those that treat powdery mildew may not be effective for downy mildew.

Fireblight is a bacterial infection and potentially very serious if not treated early. The first signs are flowers that blacken and wither; once advanced, the leaves will yellow, then blacken; ultimately, weeping cankers appear on woody stems or branches. It can affect cotoneaster.
Solution Prune out infected branches and stems immediately, cutting back to healthy growth. Burn all the material and disinfect the pruning saw to avoid infecting other plants. If the plant is too far gone, dig it up and burn it.

Honey fungus is a potentially fatal disease affecting woody plants, but can also affect perennials and bulbs, causing them to die back or lose leaves prematurely or, at worst, die.

The fungus lives on live plant material, such as tree roots, but also on dead or decaying plant matter. A sure sign of infection is the presence of honey-coloured toadstools near dying shrubs or trees. Under the bark at the base of the plant you may see black bootlace-like threads or white fungal growth with a distinct mushroomy odour; in advanced cases this will also be in the soil. Plants most likely to be affected are fruit trees, peonies, roses, forsythia and wisteria. Plants with good resistance include choisya and phlomis.
Solution There are no known remedies. Dig up and burn infected plants, removing as much of the roots as possible. Remove and destroy old rotting tree stumps and decaying leaves that can harbour the fungus. Keep plants healthy with a good watering and mulching regime to reduce their susceptibility.

Leaf spots are caused by a range of bacteria and fungi and result in spotting of leaves on a large variety of plants. The main symptoms are black spots edged with a halo of yellow of varying shapes and sizes. They may be unsightly, but do not cause serious harm.
Solution Remove all affected plant leaves by hand and water the plant at the base, rather than overhead.

Onion white rot is a soil-borne disease that can stay in the soil for years and affects ornamental alliums. The bulbs rot and are covered in a white mould, causing the plant to collapse and die.
Solution There is not much to be done but pull up the infected plants and burn them or dispose of them sealed in plastic bags. Scoop up the surrounding soil and get rid of it, as it contains fungal matter. Do not put the infected plants or soil on the compost heap as you will spread the disease. Some say rotovate the soil, but this may spread the

problem around the garden. Certainly don't plant alliums in the same soil. There are no chemical controls available.

Powdery mildew is a common fungal disease that arises from meagre growing conditions and poor air circulation. A white powdery substance on the leaves will turn them brown if untreated and result in stunted or distorted growth.
Solution Remove all infected leaves and improve air circulation by pruning. Proprietary sprays are available, but prevention is always better than cure.

Root rot is a soil-borne fungal disease that affects trees, shrubs and woody plants. Symptoms include areas of browning leaves that die back later on otherwise healthy stems. If you scrape the surface of the stem you may see blackish discolouration on the exposed tissue, which is a sure sign of infection.
Solution Preventive sprays are available for healthy plants, but once a plant is badly affected the only solution is to dig it up and burn or otherwise destroy it.

Rust is a fungal disease more likely to be seen in moist, damp conditions. It can be diagnosed by round patches of orangey brown pustules developing on the undersides of leaves. Hollyhocks (*Alcea*) and roses are susceptible.
Solution Good hygiene is the surest way to prevent rust. Prune to improve air circulation and remove and burn all infected leaves. Ensure leaves are removed from the top of the soil before they decompose and spread the infection. If rust is left unchecked, the life of the plant is at risk.

Scab is a fungal disease that primarily affects fruit, but pyracantha can also be infected. Scabby brown or grey patches appear, causing leaves to fall prematurely. It is not serious if treated quickly.
Solution Pick off by hand or prune out badly affected shoots or leaves. Burn or dispose of all infected plant material, never adding it to the compost heap, and thin branches to increase air circulation when the plant is dormant. Spray with a recommended fungicide.

Stem and **leaf rots** are bacterial infections. The first symptoms are yellowing, wilting leaves, though the base of the plant rots and decays soon afterwards.
Solution Pull up and dispose of any affected plants.

Twig blight is a fungal disease that causes blossom and flowers to wilt a couple of weeks after opening, leaving plants vulnerable to canker and shoot infections. The twigs will often die after the blossom has wilted.
Solution Cut out all affected growth in spring and again at summer and winter pruning times if still present. There are no fungicidal sprays available to the home gardener.

Verticillium wilt is caused by several types of *Verticillium* fungus. When a plant is seriously affected, the roots die and the whole plant collapses for no apparent reason. If a plant wilts inexplicably, dig it up and cut through the stems, which may reveal tell-tale brown striping.
Solution The only thing to do is dig up affected plants and burn them. There are no chemical controls available to the domestic gardener.

Propagation

There comes a time in every gardener's life when you need to increase a number of your garden plants. The easiest thing to do is to buy an identical plant at the nursery or local garden centre. However, this may not always be a practical solution. Perhaps you want to reproduce a well-loved plant that has been gifted to you; you may be moving house and would like to take cuttings of favourite shrubs to your new garden; or you may have a big garden where it is a costly exercise to buy large numbers of plants.

Propagation is the easiest and cheapest way of producing plants that are identical to those growing in your garden, and a hugely enjoyable and satisfying occupation too.

Division

Division is one of the easiest and most reliable ways to multiply perennials. It simply means dividing a plant into two or more parts. It is normally used for plants that have a clump-forming or carpeting habit and the roots can be fibrous, rhizomatous, cormed or suckering. Eremurus, echinops, euphorbias and hardy geraniums are a few of the many plants that can be multiplied by division; most herbaceous perennials are suitable for this method.

Division is most successful in early spring, when the plant is beginning to grow and there are no leaves to speak of: water loss is hugely reduced and rooting will occur more quickly. It can also be done in late autumn after flowering has finished, while the soil is still warm (warm soil encourages rapid root development). When dividing in autumn, trim the leafy areas of the plant by at least half; otherwise the roots will be overburdened by keeping the top-growth going while trying to make new roots.

Dig up an existing large clump, roots and all, and prise the rootball apart using two garden forks back to back. If it is a very large clump, repeat the process. Each new clump needs a good visible root system. Plant out in their final positions or grow on in pots to transplant later. If the roots are very fibrous and tough, use a sharp knife or cleaver to cut the sections into manageable segments. The subsequent plants will be identical in every way to the original plant.

Dividing a hardy geranium

Growing plants from seed

Many perennial plants, such as alchemilla, aquilegia and lavender, may be grown from seed and there is a huge array on offer. Growing plants from seed is cheap and relatively easy, though not always entirely reliable: some seeds germinate more easily than others, and while seed harvested from some species will grow to resemble, more or less, the plant that you collected the seed from, seed from hybrids will almost certainly produce a very different plant from the parent. Bought seed can be expected to grow true.

If you have never attempted propagation of any kind, your first results may not always be brilliant and you might have more luck with one plant than another. Plants mature according to different seasons, so sowing seed and taking cuttings are both greatly affected by changes in weather and temperature: some times of year are better than others to carry out certain tasks. However, if you follow the basic guidelines you will enjoy considerable success.

Sowing

Some seeds can be sown directly outdoors in their final flowering positions or in pots in a cold frame. Others need to be grown indoors to start with, either in a greenhouse or on a warm windowsill. Sowing seeds is easy, whether they are home-harvested or bought from seed suppliers. Seed packets are labelled with clear information on how and when to plant them, and all you need is clean pots or seed trays, fresh potting compost, a watering can and plant labels.

1 Fill your seed trays or pots with a good potting compost to 6mm/¼in or so below the rim, firm well and water lightly. A word of warning: seedlings can be affected by a process known as 'damping off', whereby all your little green seedlings can wilt and wither overnight. A light drenching of Cheshunt Compound (available from garden centres) over the composted pots or trays before sowing should prevent this rather depressing condition.

2 Check the seed packet for exact instructions on sowing particular seed. Some need to be covered with compost, while others can sit lightly on the surface. Larger seeds can be sown two or three to a pot. Label each tray or pot clearly and water lightly again. This is best done with a mister or with a watering can with a 'rose' spout, to avoid washing the seed into puddles.

3 Place the seed trays or pots on a light, non-draughty windowsill, in a sheltered porch or in a greenhouse. Some seeds need a minimum temperature to germinate effectively. (Heated propagators are inexpensive to buy and will give the optimum soil temperature.)

4 Water regularly, keeping the compost moist, but not waterlogged. Within a few weeks, seedlings will appear and you can begin to open the vents on the propagator to regulate the air temperature as they grow.

SELF-SEEDING PLANTS

Many plants will seed themselves, and this can be a great advantage as it lets you multiply any plant for nothing and you don't have to bother with the whole sowing process. However, some self-seeders are so prolific that they can become a nuisance. The only way to prevent them spreading their progeny is to deadhead the fading flowers before they scatter their seed.

Erigeron self-seeds prolifically, and will fill gaps in garden walls and between cracks in paving, so is a useful plant to let go to seed. Lady's mantle (*Alchemilla mollis*) and columbine (*Aquilegia*) are legendary self-seeders: leave the seedlings that have sited themselves in desirable places and pull up those that haven't. Some self-seeding plants produce identical offspring, but for others (such as aquilegia and foxgloves) cross-pollination can result in plants that vary in flower colour, size and habit.

RIPE SEED

Ripe seed is simply fresh seed from a mature seedpod. This can be harvested before the seedpod opens and scatters it: break off ripe seedpods, lay them on a piece of newspaper and rub them between your fingers to release the seed.

There are a few clues to indicate whether you are collecting viable, ripened seed: unripe seedpods are usually green and still fleshy, and the seeds inside may be soft and are often green or white in colour, difficult to separate from the pod or sticky; ripe seed is normally hard, yellow, black or brown in colour and falls from the seed pod easily when it is cracked open.

them in a cold frame that is left open in the daytime and then closed at night. If you are growing on a windowsill, leave the window open for a few hours each day to help them adapt to outdoor temperatures.

As the days go by, you will see the plants' foliage visibly toughen. After two to six weeks they should be sufficiently robust to brave it outdoors permanently in the places they are to flower.

Growing plants from cuttings

A cutting is a small piece cut from the stem or root of an existing plant and grown on in a separate pot to produce a new plant, exactly like the one it was taken from. There are different types of cutting, taken from different parts of the plant, depending on the type of plant or shrub you are trying to reproduce. Most perennials can be propagated from cuttings.

Growing new plants from cuttings is not difficult. Here's a list of the equipment you will need, and its preparation: hygiene is an important factor when taking cuttings, and these simple practices are all part of successful propagation.

When the seedlings are big enough to handle, select the healthiest of them and pull out the weedier-looking ones, allowing the stronger, larger ones more space to grow. As the thinned seedlings mature, the seed tray will become overcrowded, with each seedling competing for nutrients and space. It is time to pot them up separately, allowing them the luxury of their own dedicated pot so they have room to develop. This process is known as 'pricking out'. Seedlings are very fragile at this stage and need careful handling and attention: pick them up between the thumb and forefinger, using the leaves and not the stem.

Hardening off

Plants grown in protected, indoor conditions cannot be put straight into the ground outdoors without a period of 'hardening off'. This is a method of gradually acclimatising your rather spoilt, pampered plants to colder outdoor conditions over a period of weeks. Once all danger of frosts has passed, simply expose the new plants to outdoor conditions.

If you have the luxury of a greenhouse, open the vents for a couple of hours to allow colder air to circulate, gradually increasing the length of time the vents are open during the day, and closing them at night; or leave

- A clean, sharp garden knife or penknife: sterilise it by holding it over a small flame and let it cool before use.
- Fresh potting compost: don't be tempted to recycle old compost that has been lying around or has been used for growing other plants.
- Clean pots: wash all pots, trays and containers with warm soapy water, rinse and allow to dry.
- A watering can with a fine 'rose' spout, full of water.
- Hormone rooting powder (optional).
- Clean, clear plastic bags.
- Plant labels and a pen or pencil.

Choosing cutting material

The first part of the process is to remove a small piece of an existing plant, which is then turned into a cutting back at the potting shed. The cutting will eventually grow into a new plant. It is important to use a clean, sharp knife when doing this.

- Always choose cutting material from a healthy, vigorous plant: a cutting taken from a weak, sickly plant will not do well.
- Avoid any shoots affected by pests, disease or damage.
- Take cutting material from strong, non-flowering sideshoots, in different areas of the plant and avoiding the base.
- Place cutting material in a sealed plastic bag and keep in a cool place: this will help to prevent water loss, but cuttings need to be potted up almost immediately, while still fresh. They are no use once they have wilted.

My experience is that cuttings root more successfully if taken in the morning, when plants are full of water. This makes sense because once detached from the parent plant the cutting material loses its ability to feed itself from roots or any other means.

Basal stem cuttings

Basal stem cuttings are taken from the strong shoots that can be seen at the base of many perennial plants in spring when new growth begins to emerge. Achillea, geranium, gypsophila and salvia can all be propagated by this method.

Detach a short stem with new leaves attached, approximately 5–8cm/2–3in long. Trim the bottom of the cutting below a node and pot up as described below.

Softwood cuttings

Softwood cuttings are taken from a young sideshoot, just below a leaf joint or node (they are sometimes referred to as **nodal cuttings**), and normally taken in spring and early summer, when there is fresh, new growth on the plant. They have a high rooting success rate but, because the growth is young and tender, they are susceptible to bruising and need to be handled delicately. They are prone to wilting, so speed is of the essence. Buddleja, caryopteris, ceanothus, cotoneaster, wisteria and yucca are all suitable for softwood cuttings.

1 Choose a healthy stem, cut a section of it away cleanly from the original plant and then make a further clean cut in that section, just below a node or leaf joint. The final cutting should be about 5cm/2in long and have two or three pairs of leaves.
2 Fill pots with good-quality potting compost and firm well, using your fingertips. Applying a light drench of antifungal solution, such as Cheshunt Compound, helps prevent fungal disease, to which young cuttings are very susceptible.
3 Holding the cutting gently by the leaves, so as not to damage the stem, dip the cut end lightly into hormone rooting powder, tapping off any excess. This is optional, but it can accelerate rooting.
4 Using a dibber, or your fingertip, make a small hole in the surface of the compost. Still holding the cutting by the leaves, insert the cutting into the hole and firm the soil gently around the base of the stem. This will ensure it makes good contact with the compost and prevent air pockets.
5 Water well, using a fine spray, label the pot and wait for the new plantlet to form roots. This should take about three to four weeks, or less in a heated propagator providing bottom heat. Never let the pot dry out, but equally don't overwater, leaving the cutting in sodden compost, as this may cause it to rot.

Internodal cuttings are softwood cuttings taken from stems severed between two nodes or buds, which have the advantage of giving more than one cutting per stem if you would like to propagate more than one plant. They are taken in early summer from stems grown that year (not the woodier stems grown the previous season), avoiding the very soft part at the top of the stem. Honeysuckle (*Lonicera*) and ivy (*Hedera*) are ideal for internodal cuttings.

Using a clean sharp knife, cut the main stem about 5cm/2in below a leaf joint and make another cut just above the upper leaf joint, so you have a cutting approximately 10–15cm/4–6in long that looks like a clear piece of stem, with a node at the bottom that will potentially make new roots and one at the top end that will make new shoots, with two winged stems at the top, one left and one right of the stem, each having a pair of leaves. Cut away the leaves and stalk from one side of the stem, as close to the node as you can, so you are left with a stem with two leaves coming off it (removing or reducing the leaves prevents excessive water loss). Pot up as above, ensuring that the stem is covered, leaving the top of the stem and the leaves showing above the compost.

Greenwood cuttings

These are taken from the shoot tips of a plant when the stems are still young and pliable, but not too fragile, and beginning to firm up somewhat (normally early to mid-summer). They are not as delicate in handling as softwood cuttings and less prone to wilt. Box (*Buxus*), callistemon, potentilla and viburnum are all suitable for greenwood cuttings.

To take a greenwood cutting, identify a sideshoot or stem of firm but bendy growth. Detach it at the point where it joins the older plant growth. You will be left with a single main stem and several leaf shoots coming off it. Trim off the soft top shoots just above the node, to leave a cutting some 25cm/10in long with three leaf nodes. Trim any large remaining leaves to half their size with a sharp knife to prevent moisture loss. You should be left with something that resembles a single stem with two winged leaves at the top. Nick the base of the stem with a sharp knife and dip this end into hormone rooting powder, shake off any excess and insert each cutting into a suitable potting compost, deep enough that they stand upright. Water well and label.

Taking softwood cuttings from phlomis

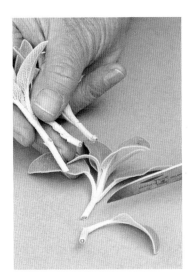

Semi-ripe cuttings

These are taken a little later in the growing season than softwood and greenwood cuttings, usually in late summer, when the stems are still young but are beginning to get firmer and buds are present, therefore they aren't so easily bruised when handling. Euonymus, elaeagnus, helichrysum, honeysuckle (*Lonicera*), ivy (*Hedera*), potentilla and St John's wort (*Hypericum*) are all suitable for semi-ripe cuttings.

Choose the current year's growth and cut a shoot that is firm at the base, but softer at the tip. Make a clean cut just below a node or leaf joint so you have a cutting approximately 10–15cm/4–6in in length. Carefully pinch out the lower leaves and the soft tip, leaving a bare stem with a pair of leaves. If the leaves are large, reduce their size by half, to minimise water loss. Pot up as above.

Stem-tip cuttings are a type of semi-ripe cutting and can be taken at any time during the growing season as long as new growth is available, using shoots with no flower buds. This type of cutting is especially good for propagating many evergreens.

To take a stem-tip cutting, select a section of stem with a healthy crown of leaves at the end, from healthy, new non-flowering shoots taken from the outer, more vigorous, part of the plant. Carefully remove the lower foliage to leave a section of bare stem and pot up as above.

Heeled semi-ripe cuttings (or heel cuttings) are also taken from semi-ripe wood but the cutting is carefully pulled away from the parent plant, rather than being cut, leaving a 'heel' or sliver of woody plant material attached to the stem which increases the chances of the successful rooting. Pot them up as described above. This is very effective for plants such as cistus and hebe.

Semi-ripe cuttings (including stem-tip and heel cuttings) are not so prone to fungal diseases as softwood cuttings, so newly potted plants can be placed in a cold frame or cool greenhouse. Providing them with bottom heat, such as a heated propagator, will encourage more rapid rooting.

Hardwood cuttings

This is one of the most successful methods of raising multiple numbers of trees or shrubs. The cutting material is taken when the plant is dormant, normally from late autumn to winter. Buddleja, cytisus, elaeagnus, potentilla, pyracantha, tamarix and viburnum are all suitable for hardwood propagation.

Taking semi-ripe cuttings from dianthus

Select a well-ripened healthy shoot that has at least three buds, about pencil thickness, from the current year's growth that has turned woody. Cut the material away from the main stem, making a clean, flush cut. Remove all leaves and make a flat cut below a bottom bud or node and a slanted cut above a top bud, trimming each stem into sections about 20–23cm/8–9in long.

If you are doing multiple cuttings, it might help to make a slit trench, about 25cm/10in deep (push your spade into the ground and then lever back and forth) and scatter a handful of sharp sand along it. Insert the cuttings vertically into the trench, with the slanted end and buds facing upward, spacing them 15cm/4in apart and leaving 5–7cm/2–3in of stem visible above the ground. Firm them in with your hands, or tread them in gently with your feet, water well and label clearly. After several months they will begin to root, and by the end of the following year they will have top-growth. Once they have lost their leaves that year, dig them up gently, roots and all, and transplant into pots or their final growing positions.

Alternatively, fill a pot with compost and sand or grit, to assist drainage, and push two thirds of the cutting into it.

Root cuttings

Taken from vigorous, young roots of a plant when dormant, usually from late autumn to winter, these are easy to do. Acanthus, border phlox, Cupid's dart (*Catananche*), poppies (*Papaver*) and verbascum can all be propagated by root cuttings.

Choose a young root close to the crown, about 10cm/4in long and as thick as a pencil. Making straight cuts at the top (nearest the crown) and angled cuts at the bottom, divide into pieces about 4cm/1½in long. Wash off the soil, pat dry and dust with a fungicide (available from garden centres). Insert each angled end of the cutting into a prepared pot of compost so that the flat ends are just visible above the surface. Cover with a thin layer of sharp sand, water, and label.

Simple layering

This is the easiest method for propagating a huge variety of plants with trailing stems and is almost always successful. Ivy (*Hedera*) and honeysuckle (*Lonicera*) are examples of the many plants that will naturally self-layer or are suitable for the layering method.

Select a healthy stem that is lying on the soil. Either simply weight it down with a stone, or bury a plastic garden pot, filled with good-quality potting compost, into the soil and peg the stem across the top. Cover lightly with compost so it is firmly anchored in the soil, label clearly, water regularly, and wait. A few months later this section of the stem will have produced roots and new shoots. Detach it from the parent plant, lift the pot if used, water it well and, hey presto – you have a new plant!

Bulbs and corms

Bulbs and corms naturally increase by forming clumps under the soil. To propagate bulbous plants which are becoming overcrowded, wait until the plant is heading into dormancy, just as foliage has died down. Dig around the parent plant and gently remove new bulblets or corms. If they are close to the size of a mature bulb they can be replanted straight away into new flowering positions. Smaller bulbs or corms can be grown on in pots until they reach a reasonable size (normally within two years) and planted out in spring or autumn.

The gardening year

Gardening is as much about good planning as anything. Get yourself organised and enjoyment flows as a matter of course. Establishing a mulching regime and growing plants that are not high maintenance will give you more time to devote to seasonal tasks without flapping that you'll never get it all done, and still leave you plenty of time for a chilled glass of something nice!

Spring

Spring is always a busy season. Most plants will appreciate feeding as they begin to grow. You can use organic mulches for plants in the ground (see page 94) or slow-release fertilisers for plants in pots. Mulching is also the first line of defence in equipping plants against drought, so taking time to mulch both your plants and the soil will pay dividends once the hot weather moves in. Remember: don't mulch soil that is dry or frozen.

A mixed purple and silver border in summer

Spring is a good time to establish new plants, transplant home-grown plants from the cold frame or start seed sowing. Spring division is very successful, so take a look at which perennials you might like to increase and divide them now, replanting into humus-rich soil to bolster their drought-busting qualities.

Walk around the garden and see if any plant needs your attention. Most drought-tolerant plants need very little care, but frosts may have lifted plants out of the soil a little, so firm them back down. Place stakes for larger plants that could do with extra support later in the year, cut back herbaceous perennials and cut out any dead, damaged or diseased material from trees and shrubs.

As the growing season gathers pace, weeds begin their relentless march. Weeding is a necessary chore, but mulching goes a long way to deterring weeds. Try to weed little but often, so you aren't confronted with a tangled infestation.

If you are growing some drought-lovers in pots, remember to mix in water-swelling granules with the compost to help them cope with drier weather later in the year. Check your water butts and seeping hoses are all functioning well, and replace faulty taps.

Summer

Summer is an easy time, when you can enjoy your garden. Keeping up with watering, weeding, deadheading and securing vigorous new growth is the most you will have to do. You may want to plunge a few bedding plants into the empty gaps that often appear in the summer border.

Pots and containers may need to be watered every other day, or once a week if using swell granules, especially in dry periods. Depending on your soil and whether you mulch the beds, as well as the

weather conditions, flower borders will need watering at least three times a week, but a well-mulched bed will usually do well with a weekly soaking. It is also an ideal time for taking softwood and greenwood cuttings.

Fading ornamental grasses touched by frosts in winter

Autumn

Autumn is a time to tidy. Clear up old leaves and garden debris to help promote a healthy, hygienic garden environment and to prevent disease overwintering in decaying matter. Enjoy your autumn-flowering plants; note the plants that did well in drought and the ones that suffered, so you know which are reliable or which will need boosting next year.

This is another good time to plant, but don't leave it too late in the year, as planting when the soil is warm will help new plant roots establish more quickly. It is an excellent time to divide plants such as border phlox and lychnis, or take root cuttings of acanthus or verbascum.

Winter

In late winter or early spring, as the beginning of the growing season approaches, prepare the garden for the months ahead. Inspect the garden for dead, diseased or damaged plants, and prune out or cut down spent, brown foliage and any dead or diseased plant matter. Wash out all your garden pots with warm soapy water and a stiff brush, and give the greenhouse (if you have one) a good hosing down both inside and out.

As the days lengthen, but the weather is still a little brisk for sitting out, make yourself useful by getting a head start on the weeding and applying garden mulches to your flowerbeds and plants. Take hardwood cuttings of shrubs such as buddleja, ceanothus and elaeagnus, as you can never have too many of a good thing, and look forward to the start of another gardening year.

Drought-tolerant plants for specific purposes

Drought-tolerant plants can fill many roles, and these categories are designed to help you choose what works best for your garden's particular conditions. If you pick a plant that has great foliage, a reasonable flowering period and good architectural form, you won't go far wrong.

Coastal areas
Aconitum napellus
Agapanthus 'Black Pantha'
Alcea rosea 'Nigra'
Armeria maritima 'Vindictive'
Atriplex halimus
Aurinia saxatilis 'Citrina'
Brachyglottis (Dunedin
 Group) 'Sunshine'
Centranthus ruber
Cistus ladanifer 'Minstrel'
Cordyline australis
 'Torbay Dazzler'
Coronilla valentina subsp.
 glauca 'Citrina'
Crocosmia 'Lucifer'
Cynara cardunculus
Cytisus × praecox
 'Warminster'
Elaeagnus × ebbingei
 'Limelight'
Eryngium × tripartitum
Erysimum 'Bowles' Mauve'
Foeniculum vulgare
 'Purpureum'
Griselinia littoralis 'Variegata'
Hebe pinguifolia 'Pagei'
Jasminum nudiflorum
Juniperus horizontalis
Lavandula angustifolia
 'Hidcote'
L. viridis
Lonicera fragrantissima
L. pileata
Lupinus arboreus
Lychnis coronaria
Ophiopogon planiscapus
 'Nigrescens'
Perovskia 'Blue Spire'
Phlomis fruticosa
Phormium tenax
Pittosporum tenuifolium
 'Silver Queen'
Rosmarinus officinalis
 Prostratus Group
Salvia argentea
Tamarix tetrandra
Ulex europeaus
Viburnum tinus 'Gwenllian'
Yucca gloriosa

Fragrance
Agastache 'Black Adder'
Allium cristophii
A. sphaerocephalon
Artemisia abrotanum
A. 'Powis Castle'
Buddleja davidii 'Black Knight'
Brugmansia aurea
Calamintha grandiflora
Callistemon citrinus
 'Splendens'
Caryopteris × clandonensis
 'Heavenly Blue'
Chamaemelum nobile
Choisya ternata
Cistus ladanifer 'Minstrel'
Cordyline australis
 'Torbay Dazzler'
Coronilla valentina subsp.
 glauca 'Citrina'
Crambe cordifolia
Cynara cardunculus
Cytisus × praecox
 'Warminster'
Deutzia × kalmiiflora
Dianthus barbatus
Elaeagnus × ebbingei
 'Limelight'
Foeniculum vulgare
 'Purpureum'
Geranium macrorrhizum
 'Ingwersen's Variety'
Helichrysum italicum
H. petiolare
Hemerocallis 'Bela Lugosi'
Jasminum nudiflorum
Juniperus horizontalis
Lavandula angustifolia
 'Hidcote'
L. stoechas
L. viridis
Lonicera fragrantissima
L. × purpusii 'Winter Beauty'
Lupinus arboreus
Mahonia aquifolium 'Apollo'
M. × media 'Winter Sun'
Nepeta 'Six Hills Giant'
Nerine bowdenii
Oenothera fruticosa
 'Fyrverkeri'

Olea europaea
Origanum vulgare
Perovskia 'Blue Spire'
Phlomis fruticosa
Phlox divaricata 'May Breeze'
Pittosporum tenuifolium
 'Silver Queen'
Rosa banksiae 'Lutea'
R. Golden Celebration
R. Summer Song
Romneya coulteri
Rosmarinus officinalis
R. o. Prostratus Group
Ruta graveolens
 'Jackman's Blue'
Salvia officinalis
 'Purpurascens'
Sarcococca hookeriana var.
 digyna 'Purple Stem'
Thymus pulegioides
 'Archer's Gold'
Ulex europeaus
Wisteria sinensis

Shade
* Suitable for full shade
Aconitum napellus
Ajuga reptans
Alstroemeria aurea
Anaphalis margaritacea
Bergenia cordifolia 'Purpurea'
Brachyglottis (Dunedin
 Group) 'Sunshine'
Buxus sempervirens *
Calamintha grandiflora
Ceratostigma willmottianum
Choisya ternata
Cordyline australis
 'Torbay Dazzler'
Cotoneaster dammeri
C. horizontalis *
Crambe cordifolia
Crocosmia 'Lucifer'
Danae racemosa *
Deutzia × kalmiiflora
Digitalis purpurea
Duchesnea indica *
Echinacea purpurea
Elaeagnus × ebbingei
 'Limelight'

Euonymus fortunei
 'Silver Queen'
Foeniculum vulgare
 'Purpureum'
Gaura lindheimeri
Geranium clarkei
 'Kashmir White'
G. endressii
G. macrorrhizum
 'Ingwersen's Variety'
G. phaeum
Hedera helix 'Glacier' *
Heuchera 'Obsidian'
Hypericum calycinum
Jasminum nudiflorum
Juniperus horizontalis
Knautia macedonica
Lespedeza thunbergii
Liriope muscari *
Lonicera fragrantissima *
L. pileata
L. × purpusii 'Winter Beauty'
Lychnis coronaria
Mahonia aquifolium 'Apollo' *
M. × media 'Winter Sun' *
Milium effusum 'Aureum'
Olea europaea
Ophiopogon planiscapus
 'Nigrescens'
Pachysandra terminalis
Phormium tenax
Pittosporum tenuifolium
 'Silver Queen'
Pyracantha 'Mohave'
Ruta graveolens
 'Jackman's Blue'
Sarcococca hookeriana var.
 digyna 'Purple Stem' *
Sedum 'Ruby Glow'
Sternbergia lutea
Uncinia rubra
Viburnum tinus 'Gwenllian' *
Vinca minor f. alba
 'Gertrude Jekyll' *
Waldsteinia ternata *

North-facing aspects
Acanthus spinosus
Aconitum napellus
Alchemilla mollis

Anaphalis margaritacea
Aquilegia alpine
Armeria maritima 'Vindictive'
Bergenia cordifolia 'Purpurea'
Briza media
Buxus sempervirens
Danae racemosa
Deutzia × kalmiiflora
Digitalis purpurea
Duchesnea indica
Elaeagnus × ebbingei 'Limelight'
Erinus alpinus
Erodium × variabile 'Roseum'
Euonymus fortunei
 'Silver Queen'
Euphorbia myrsinites
Festuca gautieri
F. glauca 'Elijah Blue'
Geranium clarkei
 'Kashmir White'
G. endressii
G. macrorrhizum
 'Ingwersen's Variety'
G. phaeum
Hebe pinguifolia 'Pagei'
Hedera helix 'Glacier'
Hypericum calycinum
Lamium maculatum
Liriope muscari
Lonicera fragrantissima
L. pileata
L. × purpusii 'Winter Beauty'
Mahonia aquifolium 'Apollo'
M. × media 'Winter Sun'
Milium effusum 'Aureum'
Origanum vulgare
Phormium tenax
Sarcococca hookeriana var.
 digyna 'Purple Stem'
Tamarix tetrandra
Ulex europeaus
Uncinia rubra
Viburnum tinus 'Gwenllian'
Vinca minor f. alba
 'Gertrude Jekyll'
Waldsteinia ternata

Ornamental grasses
Briza media
Carex comans 'Frosted Curls'
Cortaderia selloana
 'Sunningdale Silver'
Elymus hispidus
Festuca gautieri
F. glauca 'Elijah Blue'
Hordeum jubatum
Milium effusum 'Aureum'
Miscanthus sinensis
 'Silberfeder'
Ophiopogon planiscapus

 'Nigrescens'
Panicum virgatum
 'Heavy Metal'
Pennisetum orientale
 'Karley Rose'
Stipa gigantea
Uncinia rubra

Herb gardens
Foeniculum vulgare
 'Purpureum'
Helichrysum italicum
Origanum vulgare
Rosmarinus officinalis
R. o. Prostratus Group
Ruta graveolens
 'Jackman's Blue'
Salvia officinalis
 'Purpurascens'
Thymus pulegioides
 'Archer's Gold'

Naturalising
Achillea filipendulina
 'Cloth of Gold'
Allium sphaerocephalon
Briza media
Carex comans 'Frosted Curls'
Catananche caerulea
Coreopsis 'Limerock Ruby'
Digitalis purpurea
Echinacea purpurea
Echinops ritro
Foeniculum vulgare
 'Purpureum'
Gaura lindheimeri
Hemerocallis 'Bela Lugosi'
Knautia macedonica
Milium effusum 'Aureum'
Miscanthus sinensis
 'Silberfeder'
Nerine bowdenii
Oenothera fruticosa
 'Fyrverkeri'
Ophiopogon planiscapus
 'Nigrescens'
Panicum virgatum
 'Heavy Metal'
Pennisetum orientale
 'Karley Rose'
Perovskia 'Blue Spire'
Scabiosa 'Butterfly Blue'
Stipa gigantea
Uncinia rubra
Verbena bonariensis

Ground cover
Agastache 'Black Adder'
Alchemilla mollis
Anaphalis margaritacea

Anthemis punctata subsp.
 cupaniana
Aquilegia alpina
Aurinia saxatilis 'Citrina'
Ballota pseudodictamnus
Baptisia australis
Bergenia cordifolia 'Purpurea'
Briza media
Calamintha grandiflora
Carex comans 'Frosted Curls'
Ceanothus 'Blue Mound'
Chamaemelum nobile
Cotoneaster dammeri
C. horizontalis
Danae racemosa
Duchesnea indica
Erigeron karvinskianus
Erodium × variabile 'Roseum'
Euonymus fortunei
 'Silver Queen'
Festuca gautieri
F. glauca 'Elijah Blue'
Geranium clarkei
 'Kashmir White'
G. endressii
G. macrorrhizum
 'Ingwersen's Variety'
G. phaeum
Hebe pinguifolia 'Pagei'
Hedera helix 'Glacier'
Heuchera 'Obsidian'
Hordeum jubatum
Hypericum calycinum
Juniperus horizontalis
Lamium maculatum
Liriope muscari
Lonicera pileata
Mahonia aquifolium 'Apollo'
Milium effusum 'Aureum'
Ophiopogon planiscapus
 'Nigrescens'
Pachysandra terminalis
Parahebe perfoliata
Sarcococca hookeriana var.
 digyna 'Purple Stem'
Stachys byzantina
Thymus pulegioides
 'Archer's Gold'
Uncinia rubra
Vinca minor f. alba
 'Gertrude Jekyll'
Waldsteinia ternata

Climbers
Ceanothus 'Blue Mound'
Cotoneaster horizontalis
Elaeagnus × ebbingei 'Limelight'
Euonymus fortunei
 'Silver Queen'
Hedera helix 'Glacier'

Jasminum nudiflorum
Lonicera fragrantissima
L. × purpusii 'Winter Beauty'
Pyracantha 'Mohave'
Rosa banksiae 'Lutea'
R. Golden Celebration
Wisteria sinensis

Hedging
Atriplex halimus
Buxus sempervirens
Choisya ternata
Cistus ladanifer 'Minstrel'
Danae racemosa
Elaeagnus × ebbingei 'Limelight'
Griselinia littoralis 'Variegata'
Hypericum calycinum
Jasminum nudiflorum
Lavandula angustifolia
 'Hidcote'
Lonicera pileata
Mahonia aquifolium 'Apollo'
M. × media 'Winter Sun'
Pittosporum tenuifolium
 'Silver Queen'
Pyracantha 'Mohave'
Rosa banksiae 'Lutea'
Rosmarinus officinalis
 Prostratus Group
Tamarix tetrandra
Ulex europeaus
Viburnum tinus 'Gwenllian'

Architectural
Acanthus spinosus
Achillea filipendulina
 'Cloth of Gold'
Agapanthus 'Black Pantha'
Agave americana
Allium cristophii
A. sphaerocephalon
Brugmansia aurea
Buxus sempervirens
Cordyline australis
 'Torbay Dazzler'
Cortaderia selloana
 'Sunningdale Silver'
Cotoneaster horizontalis
Crambe cordifolia
Crocosmia 'Lucifer'
Cynara cardunculus
Digitalis purpurea
Echinops ritro
Eremurus stenophyllus
Eryngium × tripartitum
Euphorbia characias
 'Black Pearl'
E. c. subsp. wulfenii
E. myrsinites
Festuca gautieri

DROUGHT-TOLERANT PLANTS FOR SPECIFIC PURPOSES

12

F. glauca 'Elijah Blue'
Foeniculum vulgare
  'Purpureum'
Mahonia aquifolium 'Apollo'
M. x media 'Winter Sun'
Melianthus major
Miscanthus sinensis
  'Silberfeder'
Olea europaea
Ophiopogon planiscapus
  'Nigrescens'
Pennisetum orientale
  'Karley Rose'
Phormium tenax
Ruta graveolens
  'Jackman's Blue'
Sedum 'Ruby Glow'
Stipa gigantea
Uncinia rubra
Wisteria sinensis
Yucca gloriosa

Slopes and banks
Ajuga reptans
Anthemis punctata subsp.
  cupaniana
Aurinia saxatilis 'Citrina'
Bergenia cordifolia 'Purpurea'
Briza media
Ceanothus 'Blue Mound'
Cerastium tomentosum
Cistus ladanifer 'Minstrel'
Cotoneaster dammeri
Crocosmia 'Lucifer'
Duchesnea indica
Erigeron karvinskianus
Erodium x variabile 'Roseum'
Euphorbia myrsinites
Geranium endressii
Helianthemum
  'Wisley Primrose'
Hypericum calycinum
Juniperus horizontalis
Lespedeza thunbergii
Liriope muscari
Pachysandra terminalis
Parahebe perfoliata
Rosmarinus officinalis
R. o. Prostratus Group
Sarcococca hookeriana var.
  digyna 'Purple Stem'
Vinca minor f. *alba*
  'Gertrude Jekyll'
Waldsteinia ternata

Walls/paved areas
Brugmansia aurea
Cerastium tomentosum
Cotoneaster horizontalis
Erigeron karvinskianus

Erinus alpinus
Euonymus fortunei
  'Silver Queen'
Jasminum nudiflorum
Lonicera fragrantissima
L. x purpusii 'Winter Beauty'
Origanum vulgare
Pittosporum tenuifolium
  'Silver Queen'
Romneya coulteri
Rosa banksiae 'Lutea'
Rosmarinus officinalis
  Prostratus Group
Sempervivum
  'Commander Hay'
Thymus pulegioides
  'Archer's Gold'

Containers
Agapanthus 'Black Pantha'
Alstroemeria aurea
Aurinia saxatilis 'Citrina'
Ballota pseudodictamnus
Briza media
Brugmansia aurea
Buxus sempervirens
Callistemon citrinus
  'Splendens'
Carex comans
  'Frosted Curls'
Cordyline australis
  'Torbay Dazzler'
Coronilla valentina subsp.
  glauca 'Citrina'
Cotoneaster horizontalis
Dianthus barbatus
Elymus hispidus
Erigeron karvinskianus
Erinus alpinus
Erodium x variabile 'Roseum'
Euphorbia myrsinites
Festuca gautieri
F. glauca 'Elijah Blue'
Hebe pinguifolia 'Pagei'
Helichrysum petiolare
Heuchera 'Obsidian'
Juniperus horizontalis
Lavandula stoechas
L. viridis
Liriope muscari
Lonicera x purpusii
  'Winter Beauty'
Miscanthus sinensis
  'Silberfeder'
Nerine bowdenii
Olea europaea
Ophiopogon planiscapus
  'Nigrescens'
Origanum vulgare
Panicum virgatum

'Heavy Metal'
Pennisetum orientale
  'Karley Rose'
Phlox douglasii
  'Boothman's Variety'
Phormium tenax
Potentilla x tonguei
Rosa Golden Celebration
R. Summer Song
Ruta graveolens
  'Jackman's Blue'
Sarcococca hookeriana var.
  digyna 'Purple Stem'
Sempervivum
  'Commander Hay'
Sternbergia lutea
Stipa gigantea
Thymus pulegioides
  'Archer's Gold'
Uncinia rubra
Yucca gloriosa
Zauschneria californica
  'Dublin'

Silver-leaved
Anaphalis margaritacea
Anthemis punctata subsp.
  cupaniana
Artemisia abrotanum
A. 'Powis Castle'
Atriplex halimus
Brachyglottis (Dunedin
  Group) 'Sunshine'
Carex comans 'Frosted Curls'
Catananche caerulea
Cerastium tomentosum
Convolvulus cneorum
Cynara cardunculus
Echinops ritro x
Elymus hispidus
Festuca glauca 'Elijah Blue'
Helichrysum italicum
H. petiolare
Juniperus horizontalis
Lavandula angustifolia
  'Hidcote'
L. stoechas
Lychnis coronaria
Rosmarinus officinalis
R. o. Prostratus Group
Salvia argentea
Stachys byzantina

Foliage
Acanthus spinosus
Agastache 'Black Adder'
Agave americana
Ajuga reptans
Aquilegia alpina
Artemisia abrotanum

A. 'Powis Castle'
Atriplex halimus
Ballota pseudodictamnus
Bergenia cordifolia 'Purpurea'
Brachyglottis (Dunedin
  Group) 'Sunshine'
Buxus sempervirens
Calamintha grandiflora
Carex comans
  'Frosted Curls'
Cordyline australis
  'Torbay Dazzler'
Cortaderia selloana
  'Sunningdale Silver'
Cotoneaster dammeri
Crocosmia 'Lucifer'
Danae racemosa
Elaeagnus x ebbingei
  'Limelight'
Elymus hispidus
Euonymus fortunei
  'Silver Queen'
Euphorbia characias
  'Black Pearl'
E. myrsinites
Festuca gautieri
F. glauca 'Elijah Blue'
Foeniculum vulgare
  'Purpureum'
Geranium clarkei
  'Kashmir White'
Griselinia littoralis 'Variegata'
Hebe pinguifolia 'Pagei'
Hedera helix 'Glacier'
Helichrysum italicum
H. petiolare
Hemerocallis 'Bela Lugosi'
Heuchera 'Obsidian'
Juniperus horizontalis
Lamium maculatum
Lavandula angustifolia
  'Hidcote'
L. stoechas
L. viridis
Lonicera pileata
Lupinus arboreus
Melianthus major
Milium effusum 'Aureum'
Miscanthus sinensis
  'Silberfeder'
Olea europaea
Ophiopogon planiscapus
  'Nigrescens'
Pachysandra terminalis
Panicum virgatum
  'Heavy Metal'
Pennisetum orientale
  'Karley Rose'
Phormium tenax
Pittosporum tenuifolium

'Silver Queen'
Pyracantha 'Mohave'
Ruta graveolens
 'Jackman's Blue'
Salvia argentea
Sarcococca hookeriana var.
 digyna 'Purple Stem'
Sempervivum
 'Commander Hay'
Uncinia rubra
Viburnum tinus 'Gwenllian'
Waldsteinia ternata

Berries/fruits
Cordyline australis
 'Torbay Dazzler'
Cotoneaster dammeri
C. horizontalis
Danae racemosa
Duchesnea indica
Euonymus fortunei
 'Silver Queen'
Griselinia littoralis 'Variegata'
Hedera helix 'Glacier'
Juniperus horizontalis
Lonicera fragrantissima
L. pileata
Mahonia aquifolium 'Apollo'
M. × media 'Winter Sun'
Olea europaea
Ophiopogon planiscapus
 'Nigrescens'
Pyracantha 'Mohave'
Sarcococca hookeriana var.
 digyna 'Purple Stem'
Viburnum tinus 'Gwenllian'

Blue flowers
Ajuga reptans
Aquilegia alpina
Aster turbinellus
Baptisia australis
Caryopteris × clandonensis
 'Heavenly Blue'
Catananche caerulea
Ceanothus 'Blue Mound'
Ceratostigma willmottianum
Echinops ritro
Erinus alpinus
Eryngium × tripartum
Lavandula angustifolia
 'Hidcote'
Nepeta 'Six Hills Giant'
Parahebe perfoliata
Perovskia 'Blue Spire'
Phlox divaricata 'May Breeze'
Rosmarinus officinalis
R. o. Prostratus Group
Scabiosa 'Butterfly Blue'

Green flowers
Euonymus fortunei
 'Silver Queen'
Euphorbia characias
 'Black Pearl'
E. c. subsp. wulfenii
Griselinia littoralis 'Variegata'

Mauve/purple flowers
Agapanthus 'Black Pantha'
Agastache 'Black Adder'
Aconitum napellus
Allium cristophii
A. sphaerocephalon
Alcea rosea 'Nigra'
Baptisia australis
Bergenia cordifolia 'Purpurea'
Buddleja davidii 'Black Knight'
Cynara cardunculus
Dianthus barbatus
Digitalis purpurea
Erinus alpinus
Erysimum 'Bowles' Mauve'
Geranium phaeum
Hemerocallis 'Bela Lugosi'
Knautia macedonica
Lavandula stoechas
Lespedeza thunbergii
Liriope muscari
Ophiopogon planiscapus
 'Nigrescens'
Phlox douglasii
 'Boothman's Variety'
Pittosporum tenuifolium
 'Silver Queen'
Salvia officinalis
 'Purpurascens'
Stachys byzantina
Verbena bonariensis
Thymus pulegioides
 'Archer's Gold'

Yellow flowers
Achillea filipendulina
 'Cloth of Gold'
Alchemilla mollis
Alstroemeria aurea
Artemisia abrotanum
A. 'Powis Castle'
Aurinia saxatilis 'Citrina'
Brachyglottis (Dunedin
 Group) 'Sunshine'
Brugmansia aurea
Coronilla valentina subsp.
 glauca 'Citrina'
Cytisus × praecox
 'Warminster'
Duchesnea indica
Eremurus stenophyllus
Euphorbia myrsinites

Foeniculum vulgare
 'Purpureum'
Helianthemum
 'Wisley Primrose'
Helichrysum italicum
Hypericum calycinum
Jasminum nudiflorum
Lavandula viridis
Lupinus arboreus
Mahonia aquifolium 'Apollo'
M. × media 'Winter Sun'
Milium effusum 'Aureum'
Oenothera fruticosa
 'Fyrverkeri'
Phlomis fruticosa
Rosa banksiae 'Lutea'
R. Golden Celebration
Ruta graveolens
 'Jackman's Blue'
Sisyrinchium striatum
Sternbergia lutea
Ulex europeaus
Waldsteinia ternata

Red flowers
Callistemon citrinus
 'Splendens'
Coreopsis 'Limerock Ruby'
Crocosmia 'Lucifer'
Melianthus major
Phormium tenax
Sedum 'Ruby Glow'
Zauschneria californica
 'Dublin'

Pink flowers
Armeria maritima 'Vindictive'
Bergenia cordifolia 'Purpurea'
Calamintha grandiflora
Centranthus ruber
Cotoneaster horizontalis
Crepis incana
Deutzia × kalmiiflora
Dianthus barbatus
Echinacea purpurea
Erinus alpinus
Erodium × variabile
 'Roseum'
Geranium endressii
G. macrorrhizum
 'Ingwersen's Variety'
Lamium maculatum
Lychnis coronaria
Nerine bowdenii
Origanum vulgare
Salvia argentea
Sarcococca hookeriana var.
 digyna 'Purple Stem'
Tamarix tetrandra

Orange flowers
Papaver orientale 'Allegro'
Potentilla × tonguei
Rosa Summer Song
Verbascum 'Helen Johnson'

White flowers
Anaphalis margaritacea
Anthemis punctata subsp.
 cupaniana
Atriplex halimus
Ballota pseudodictamnus
Cerastium tomentosum
Chamaemelum nobile
Choisya ternata
Cistus ladanifer 'Minstrel'
Convolvulus cneorum
Cordyline australis
 'Torbay Dazzler'
Cortaderia selloana
 'Sunningdale Silver'
Cotoneaster dammeri
Crambe cordifolia
Deutzia × kalmiiflora
Elaeagnus × ebbingei
 'Limelight'
Erigeron karvinskianus
Gaura lindheimeri
Geranium clarkei
 'Kashmir White'
Gypsophila paniculata
Hebe pinguifolia 'Pagei'
Helichrysum petiolare
Heuchera 'Obsidian'
Libertia grandiflora
Lonicera fragrantissima
L. pileata
L. × purpusii 'Winter Beauty'
Olea europaea
Pachysandra terminalis
Pyracantha 'Mohave'
Romneya coulteri
Salvia argentea
Sarcococca hookeriana var.
 digyna 'Purple Stem'
Viburnum tinus 'Gwenllian'
Vinca minor f. alba
 'Gertrude Jekyll'
Yucca gloriosa

Plant index